HOOD COUNTY

AF580809

COMPUTERS IN SPACE

ALLEN L. WOLD

A COMPUTER APPLICATIONS BOOK
FRANKLIN WATTS ■ 1984
NEW YORK ■ LONDON ■ TORONTO ■ SYDNEY

Photographs courtesy of:
NASA: pp. 6, 22, 28, 35, 36, 47, 48, 54, 61, 69, 77, 84 87, 90, 99; Kitt Peak National Observatory: pp. 19, 115; The National Radio Astronomy Observatory: pp. 27, 42, 112; Smithsonian Instution, Whipple Observatory: p. 41.

Library of Congress Cataloging in Publication Data

Wold, Allen L.
Computers in space.

(A Computer applications book)
Includes index.
Summary: Discusses how computers aid our research into the nature of the world, the sun and solar system, the stars, the galaxies, and the cosmos.
1. Astronomy—Data processing—Juvenile literature.
2. Space sciences—Data processing—Juvenile literature.
[1. Astronomy—Data processing. 2. Space sciences—Data processing] I. Title. II. Series.
QB51.3.E43W65 1984 520'.285'4 84-5251
ISBN 0-531-04847-0

Printed in the United States of America
5 4 3 2 1

CONTENTS

COMPUTERS IN SPACE

CHAPTER 1

TELESCOPES, ROCKETS, AND COMPUTERS

Human beings have always looked toward the heavens with interest. The sky presents us with the most dramatic of changes, from day to night, clear weather to stormy. The sun not only lets us see, it powers the whole of the biosphere on earth. The stars provide us with myths, and with clocks and calendars. We study the universe because it is fascinating in itself, and because we want to know more about our place in it.

Our interest in space has not always been scientific. Our ancestors looked at the sky, the sun and moon, the planets, and the stars, and there sought divine, mystical, or spiritual guidance, explanations, and meaning. Some of us still do this today. We also find beauty, mystery, and wonder in the heavens.

Though most of us no longer believe that the mythological gods live beyond the sky, or that burning incense will change the weather, our superstitious interest in the sky cannot be completely separated from our scientific study of it. Humans watched the stars, saw them rise and set at different times, and noticed that they changed their positions in an annual cycle. The moon changed its phases, rose higher or lower in the sky at different seasons, and seemed somehow connected with the tides. The sun provided not only heat and light but also kept plants alive and green.

These observations led to both astrology and astronomy, too much the same enterprise at first to be separated. Astrology deals with the supposed meanings and influences of the heavenly bodies and attempts to predict the future. Astronomy, on the other hand, deals with the scientific understanding of the origin and evolution of the universe.

As we learned more about the true nature of those lights in the sky, it became evident that although Mars in conjunction with the moon might or might not have an influence on someone's personality, the influence of the moon on the tides was clear and unarguable. The cycles of the stars and sun were soon tied to the cycles of the seasons and to the growth of crops.

In addition, almost from the very beginning, humans used calculation devices to help them in their observations of the heavens. Among the oldest artifacts of humankind are monolithic structures, circles of stones, or holes in the ground where poles were once set. Many anthropologists and archaeologists believe these structures were used to predict certain astronomical events. In a way, then, one could look at Stonehenge as a kind of analog computer, designed to predict the equinoxes, solstices, and certain significant positions of the moon.

Computers and space science go back a long way together.

* * * * * * * *

Today, almost all of space science depends on computers in some way or other. This dependence is so profound and so all-encompassing that it is hard to talk about space science without at least tacitly acknowledging the computer's role. But popular books and articles on astronomy, telescopy, rocketry, or voyages to the planets seldom mention the computer, though it is usually there somewhere, helping in the design, manufacture, and control of telescopes and spacecraft, solving mathematical problems, creating pictures, and making communications possible.

Without the computer, our observations would be limited to what we could achieve using just our eyes, lenses and mirrors, and photographic film. Our explorations could not go far beyond our atmosphere, even with rockets. Perhaps no area of human research and speculation is as dependent on computers as is the collection of sciences, technologies, and theories we call space science.

In turn, space science has been at least in part responsible for the development of the modern computer. As the science of rocketry grew, the need for onboard control, as opposed to remote control, became ever more evident. More than that, simple mechanical or clockwork control soon proved inadequate. What was needed was a device that functioned more like the human mind.

Computers built of vacuum tubes and relays, requiring vast amounts of power, could not possibly be put on board even today's spacecraft, let alone on the tiny rockets of the early years of space exploration. The size of computers came down, in part, because smaller was cheaper, faster, more efficient, and more reliable, but also because computers small enough to fit were needed for the onboard guidance and control of our spacecraft.

Without the need for such computers—and other electronic components that had to be made as small and light as possible—modern computers might still be relying on large, old-fashioned transistors, and the home computer would probably still be decades in the future. As it is, anyone today can own a computer more powerful than the

Mark I, the first computer built by IBM, or the ENIAC, built in 1946 by the University of Pennsylvania under contract to the Army.

How space science has affected the development of computers is a story in itself, and we will say little about it here, except in passing. Our concern in this book is mainly with how computers aid us in our research into the nature of our world, our sun and solar system, the stars, the galaxies, and the cosmos.

* * * * * * * *

Space science can be divided into roughly two areas: observation (looking at something) and exploration (going there). These two areas are supplemented by theory, modeling, and simulation. Space science also involves the design and manufacture of instruments.

OBSERVATION

Observation is by far the oldest activity in space science. Indeed, in its most elemental form, all it needs is a pair of eyes, to look up at the night sky. But the earliest astronomers soon discovered that their observations were more accurate and reliable with mechanical help. They invented devices that could keep track of the passage of time and that could point to particular places in the sky at special times. If indeed Stonehenge was a primitive form of observatory, then that was what it was used for.

A few pretelescopic observatories had some rather sophisticated equipment. Tycho Brahe, perhaps the greatest astronomical observer before the invention of the telescope, built large numbers of devices that were used to locate stars in the sky, measure the angular separation between the stars or planets, and calculate the timing of astronomical events. These instruments, which were very accurate, and the observational methods that Brahe developed allowed him to prove that a comet was farther away from the earth than the moon was, a fact that was not generally believed possible at that time.

But there is a limit to what can be observed without a telescope. Galileo used one of the first telescopes to look at the sky, and went on to improve the instrument. From this first simple device has evolved a large number of different kinds of sophisticated optical instruments—from refracting telescopes with complex arrangements of lenses and reflecting telescopes with single mirrors to the Multiple Mirror Telescope at Kitt Peak, the Russian 236-inch (6-m) reflector in the Caucasus, and the Space Telescope.

But visible light is not the only portion of the electromagnetic spectrum that can be detected and magnified. Ultraviolet light, infrared radiation, radio waves, X-rays, and even cosmic rays (which are not radiation at all but very fast-moving particles) can be used to observe planets, stars, and the space between them. Most of the devices that can detect these forms of radiation, as well as many modern optical telescopes, depend on the computer for their control. Computers are also used to convert the data produced by these detectors into pictures or other forms of information that we can understand.

But we are not content just to look. We also want to visit the places we see from afar. This is the second major area of space science—exploration.

EXPLORATION

One of the most exciting endeavors of modern scientific research and development has been the exploration of space, which began with hot-air balloons and has progressed to sounding rockets, orbital satellites, Martian landers, and spacecraft probes that are now traveling beyond the solar system, heading for the stars. The science of rocketry is what makes all our explorations possible. The most sophisticated piece of experimental equip-

ment does no good if we can't get it to go where we wish.

The first rockets were invented in China in the thirteenth century. At first, they were little more than toys and later were used primarily for military purposes.

Two-stage solid-fuel rockets were invented in the 1850s as a means of carrying lifelines to victims of ship disasters. Robert Goddard, inspired by a dream of traveling to Mars, developed the liquid-fuel rocket, and from that came the Titans, Atlases, Saturns, space shuttles, and other rockets that today boost satellites and people into orbit.

But the more sophisticated the rocket, and the farther it has to travel, the less it can depend on simple aiming, clockwork guidance systems, or even remote control. Onboard computers are essential. Without them, rocketry would be limited to military barrage and atmospheric sounding.

A CLOSER LOOK AT COMPUTERS

The word "digital," as in *digital computer,* comes from the word for "finger." Our fingers were our first counting and calculating tools. But, as portable as they are, fingers are not very efficient, have very little memory, and are limited in their ability to calculate large sums.

Robert Goddard, who is considered the "Father of American Rocketry," is shown here with the first liquid-propellant rocket, which achieved flight on March 16, 1926.

After fingers came the counting board, which resembled an abacus without wires, and then came the abacus itself. Blaise Pascal, in 1642, invented a mechanical calculator that could add and subtract. The use of gears and teeth to manipulate numbers culminated in Charles Babbage's designs for a "Difference Engine" and an "Analytical Engine," neither of which were ever completed. The concepts behind their designs were sound, but the problems of manufacture were insurmountable at the time.

The Analytical Engine was designed to be programmed and even to have a mechanical memory. But it wasn't until the development of electronic devices that a true computer could be built, first using electromechanical relays, then vacuum tubes, then transistors, and now very large-scale integration (VLSI) of thousands of microscopic transistors on silicon chips.

Another aspect in the development of computers involved what they could do. The simplest operation is counting. Calculation involves the manipulation of numbers, and pocket calculators do just that, one step at a time. When a series of different operations or steps can be performed automatically, then you have a true computer, whether it is mechanical and steam-powered or electronic and solar-powered. Today's programmable calculators are really just small computers, and the distinction between them and what are called pocket computers is rapidly disappearing.

Basically, computers do only two things, though sophisticated design and clever programming can enhance these two operations and even add to them. The two things are add and branch. Almost any mathematical operation can be reduced to a series of additions, and a computer can perform hundreds, thousands, or even millions of these additions a second. This is what gives it the power to calculate and compute.

Branching, which is equally important, allows the computer to make choices. In its simplest form, branching is this: If an input is 1, then do this; if it is 0, then do that. And that is the basis for all computer control, whether of microwave ovens or Martian landers. These two abilities, combined with a stored program, are what make computers useful.

DATA MANIPULATION AND ANALYSIS

The purpose of a spacecraft or probe sent out beyond the confines of the earth's atmosphere is to observe, perhaps small bodies relatively close at hand, such as a solar-system moon, or perhaps larger structures farther away, such as distant galaxies or quasars. As each of these probes, whatever its nature, makes its observations, it produces vast amounts of data that come back to earth in huge streams.

This data, which is binary—a series of 0's and 1's—must be converted into numbers reflecting the values observed. These values can be examined as they are, but the amount of data is sometimes so great that doing so makes little sense. When the *Mariners* went by Mars in August 1969, the engineers at the Jet Propulsion Laboratory in Pasadena, California, had been able to add to the equipment an experimental device that increased the rate of data transmission from 8⅓ bits per second to an effective rate of 16,200 bits per second. The rate at which the data was recorded by the spacecraft was an effective 117,000 bits per second.

A half-hour's worth of data could be transmitted in about three hours. A printout of this data would make a stack of paper nearly 3 feet (.9 m) high. Needless to say, no person could make sense out of such a mass of information. It had to be handled by a computer, to be converted into forms people could understand, such as pictures. The resolution, or quality, of the pictures of each scene wasn't always perfect, so several pictures of the same area were combined, digitally, by computer, to compensate for this and for other problems, such as the occasional loss of a bit or two in transmission. This was just an experiment, and the regular, slow system of transmission was still available, though not used.

COMPUTERS, SPACE SCIENCE, AND YOU

Space science today is showing that even knowledge about things light-years away in space and millions of years away in time can improve our lives in practical ways, as well as provide us with intellectual satisfaction.

In science, we try to determine the nature of the universe, or small parts of it; to trace the history of the universe, or its parts, up to the present from different times in the past; and to predict the future. After that, practical applications follow, such as where to drill for oil, how to predict earthquakes, or how to derive power from exotic sources.

Not to be discounted are philosophical questions, such as, what is the meaning of life? How and why do we exist? Where might we be going? Are we alone? and so on. Indeed, it is the search for answers to these and similar questions that has driven much of the practical efforts of humankind through the ages.

Computer science, on the other hand, deals with the very small and the very near—including our own thought processes. As we shall try to show, computer science and space science are closely interrelated. Space science is perhaps the greater beneficiary of this symbiosis, though computers have benefited, too. For example, gallium arsenide, when used in computer chips, is much better than silicon in terms of number of circuits that can be etched onto it, speed of electron flow, and reliability. This substance is very hard to make on earth but not so hard in the gravity-free conditions of orbit. On earth, gravity-induced convection currents stir up impurities in the gallium arsenide, rendering more than 90 percent of the newly made chips useless. In space, with no gravity, this ratio could be reversed.

In the chapters to follow, we will try to indicate some of the scope of computer use in learning about our universe. It would be futile to try to list all the possible space-oriented applications, especially since new fields, new discoveries, and new technologies would make any such listing out of date before it was published. But there are, broadly speaking, three general areas where interest in computers and astronomy overlap.

The first involves designing computers for space applications. Those computers that go into satellites and spacecraft, though small, are not like the desktop computers with which we are all familiar. They are more like single-board computers (just one circuit board with all the necessary electronic components but no keyboard or screen), or "black boxes." Some of these are very simple and small; others are more complex and considerably larger. All of them, whether they are adapted from existing computers or are based on completely new designs, have to be made to accommodate the restrictions on mass, temperature range, stress, lack of atmosphere, and so on.

The large computers that stay on the ground must also be carefully designed. Here there is more leeway, however, a greater opportunity to use existing machines adapted to the purpose.

This leads to the second area, which involves designing special peripherals, such as storage and communication devices, and constructing the interfaces that are necessary for the computer to be able to communicate with them. Special tape recorders had to be used on some of the interplanetary missions, though they were known to be relatively unreliable. Interfacing those recorders was a special problem. The Mars soil sampler was a peripheral in the same way a printer is. It received its instructions and reported back on its status. Every experiment performed by a computer in space is a special device that depends to a greater or lesser degree on computer science for control, acquisition of data, analysis, and so on.

Third, there is the designing of those applications themselves—what the experiment is supposed to do. The peripherals and interfaces can be designed only after the objectives and goals of the mission have been defined. This is only marginally computer science, but the writing of programs, for use on the ground or in space, is not. Computer programming, using a variety of sophisticated languages, little known outside the field, is perhaps the key element in a mission's success. It takes only one bug in a program to cause a rocket to crash.

Telescopes, sensors, control devices, electronic and mechanical equipment—all produce tons of data. Some of this data is handled on the spot, in "real time," especially when the data has to do with controlling the device that is providing that data. The rest of the data is handled at a later time, and many times, especially with satellites and spacecraft, at another place. All the data, however, is handled by computers and involves data processing, mathematical and statistical analysis, and perhaps imaging as well, which is one of the possible results of analysis. The programs that perform these tasks all have to be written by somebody, and they are a real challenge.

* * * * * * * *

It should be clear that, no matter what branch of space science you are interested in, a good knowledge of computer science will help, and if you intend to become a computer scientist, some of the most fascinating applications of your field lie in the realms of astronomy and related subjects.

So think about it. Whether you prefer to design computers, create the devices whereby they communicate with and affect the world, or write the programs that give them that capability, there is something for you here. As a space scientist or a computer scientist, you will find yourself involved in one of the most rewarding and challenging of professions.

CHAPTER 2

OBSERVING THE SKY FROM EARTH

Amateur astronomers are important to the whole astronomical community.. Amateurs have long made their own refracting and reflecting telescopes, either from kits or by grinding their own lenses and mirrors. Even radio telescopes, and sensors of the more exotic regions of the electromagnetic spectrum, can be homemade, following well-known theories and principles of construction.

Indeed, amateur telescopy of all kinds provides meaningful information for use by science professionals in government, industry, and research universities. Certain events—such as the appearance of a comet in the sky, or the passing of the moon between us and a star or planet (called an occultation)—can be observed with just a good pair of binoculars and a stopwatch. The information obtained from small telescopes, whether privately owned or at small colleges, is always useful.

But there is a limit to what can be done with small telescopes, even with the best of intentions. Astronomers, who are continually seeking ways of improving their ability to see the stars, planets, and galaxies, have thus constructed complex systems of mirrors and lenses to help them gather visible light too faint to be seen. They have developed ways of detecting forms of radiation in the invisible portions of the spectrum and have put their observatories into orbit around the earth above the atmosphere. To see to the edges of space, they have built sophisticated machinery, constructed with extreme precision and controlled with a delicacy unachievable by human hands. They have improved on the eye by taking pictures either on film or by use of microelectronics. Then, once the images are collected, astronomers using computers can manipulate and enhance those images in surprising ways, to make sense of confusing ones and to reveal even more information and detail. And, as observational equipment gets larger, more sophisticated, and more complicated, its design and construction become ever more demanding of precision, efficiency, and flexibility. As invisible regions of the spectrum are sampled and analyzed, and as the location of telescopes migrates from backyards to mountaintops to orbit, the need for precise control, communication, and sophisticated analysis become ever more important.

GATHERING STARLIGHT

Observation of the sky is the oldest form of space science. In spite of more dramatic hardware, such as space probes, satellites, and space shuttles, telescopes are still very important. And though other portions of the spectrum are now routinely sampled, observation in visible light is still extensive.

The usual way of recording starlight is by means of photographic film. The original way to observe the stars by telescope, with the astronomer's eye at the instrument's eyepiece, is becoming less and less common these days, except for amateurs, and even they are using other kinds of information-gathering equipment. Of course, if a telescope is under manual control, professional astronomers may still have to look through a sighting scope. But the information astronomers seek is usually captured on film—photographs of the stars and planets—because film exposed over many hours can capture far more light than a human eye can on a moment-to-moment basis.

In order to produce a good photographic image, the telescope must be aimed at its target for a rather long period of time. In many cases, the object to be observed is otherwise invisible and can be detected only after a photographic plate has been exposed for several hours. Because of the movement of the earth, keeping the telescope properly aimed by clockwork or manual control is difficult at best. But with the computer's ability to control precisely the direction of the telescope and to compensate for the earth's rotation, we are able to make observations that would otherwise be impossible to do.

This ability to aim a telescope (or radio antenna or other device) at an invisible object with extreme accuracy over long periods of time becomes even more important in keeping track of artificial satellites, which are too small to be seen by the naked eye and which move quickly. Part of this control is achieved by knowing in advance where the satellite or other astronomical object is going to be and simply pointing the telescope there.

Another way for a computer to "lock on" to the image of the star being observed (or onto a "guide star" if the star to be photographed is too faint) is to detect its drift across the observational field, and then to direct the telescope to move just a bit until the image is back where it is supposed to be. Indeed, this is the same kind of feedback system we use when looking through a small telescope, with our hands on the knobs. But the computer, making use of sophisticated imaging techniques, can do it far better.

Keeping an image in the center of the field of view requires precise control of a rather unwieldy mechanical device, the telescope, which may weigh many tons. For nearby objects, or those requiring a short exposure time, minor vibrations have little or no effect on the quality of the image produced.

But for objects that are farther away, or for those that require that the telescope be aimed at them for several hours at a time, the slightest shudder, if it persists, can distort the image, or blur it so badly that no sense can be made of it. The higher the resolution desired, or the more distant the object, the more important is the control of the telescope. Indeed, control of a telescope, and of its sheltering dome, is so important that a whole computer language was developed just for that purpose!

Charles Moore, who could program in nearly a dozen computer languages, found that none of them did exactly what he wanted them to do. He wanted them to track artifical satellites, but the languages were clumsy at the problem of process control, that is, getting a machine to do something, especially in concert with certain events. During the sixties, he made use of his knowledge of many languages to invent a new one, which he called FORTH.

Today, this is the language of choice for astronomical observatories around the world. It has since been found useful for other applications, of course, including color graphics and systems design. But as a tool for the observational astronomer, it has yet to be bettered.

COMPUTER PROCESSING OF PHOTOGRAPHS

Photographs enhance our ability to see the stars. Our eyes, as sensitive as they are, can react to the light only as it is coming in. A photographic plate, on the other hand, can accumulate light over a long period of time, and hence enables us to see objects that would otherwise be invisible to us, even if we looked through the telescope with our eyes. But because of these long exposure times, pictures can be somewhat blurred, by movement of the telescope or of the atmosphere.

Computer processing of images is done even when there are no problems with the picture. Image-enhancement techniques can be used to work with photographs that are otherwise perfectly exposed but which, in themselves, do not reveal as much about the observed object as they might.

There are a number of ways a computer can be used to correct, enhance, or otherwise improve an existing photograph. Many of these techniques were not originally intended for astronomical applications, but they have since been found useful.

A photograph is an analog representation of a scene. That is, the degree of darkness or lightness in a photograph is directly related to the amount of light that was reflected by the object when it was being viewed. In order for the computer to work with the image, the information contained in the photograph must be digitized, that is,

Two images of galaxy NGC 4736. The top photo was developed in the traditional way, but in the bottom photo, computer enhancement techniques were used to show only the ionized gas after removing the starlight from the galaxy's bulge.

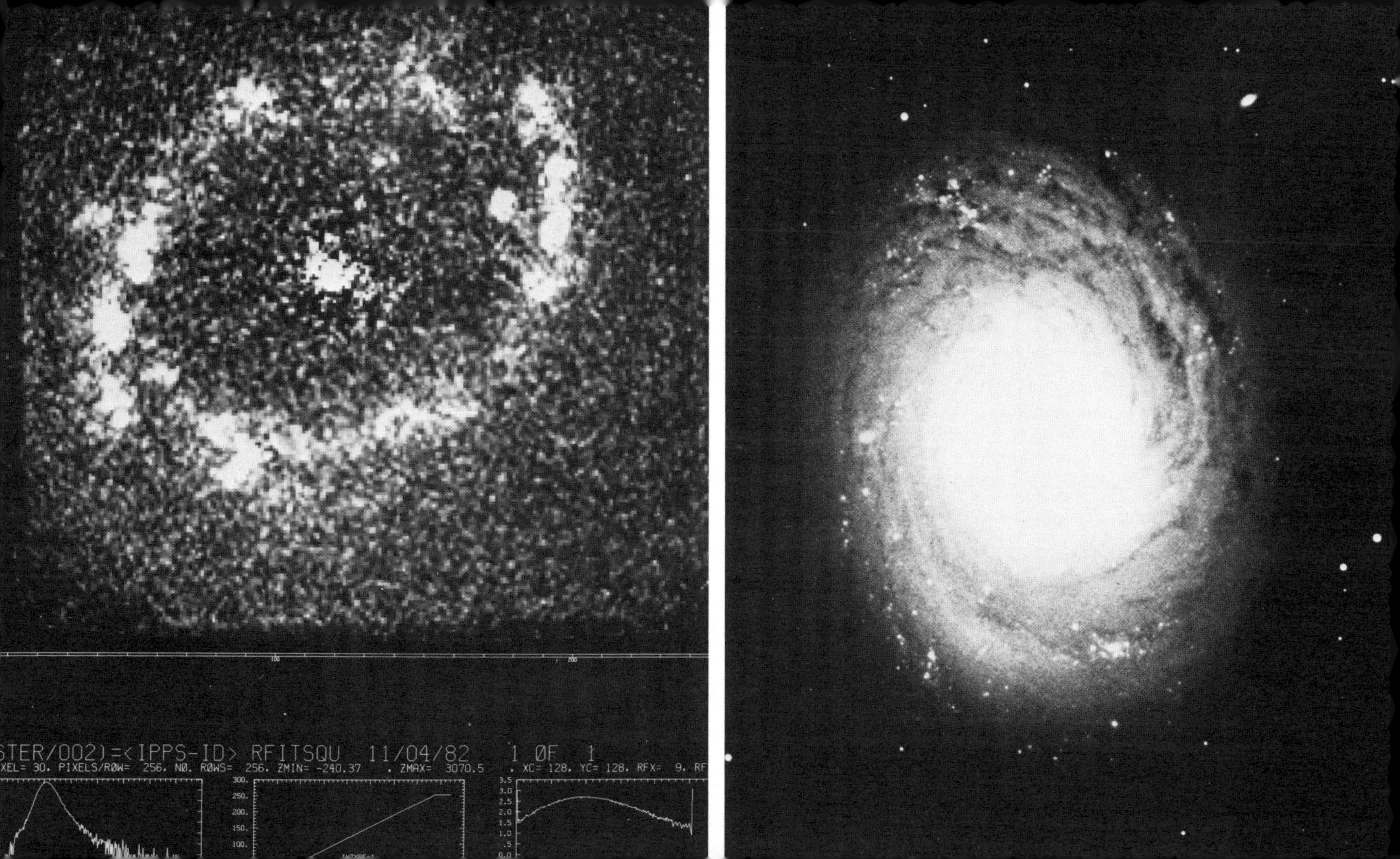

STER/002)=<IPPS-ID> RFITSQU 11/04/82 1 ØF 1
XEL= 30. PIXELS/RØW= 256. NØ. RØWS= 256. ZMIN= -240.37 , ZMAX= 3070.5 , XC= 128. YC= 128. RFX= 9. RF

reduced to numbers. One way is to scan the photograph, looking only at small sections (called picture elements, or pixels for short) at a time and recording the brightness of each of these sections. This can be done by using a device called a scanning microdensitometer, which is similar to a TV camera but more sophisticated, or a flying-spot scanner. In each of these devices, light shining through or reflecting off the picture is detected and measured one spot at a time.

A second method uses charge-coupled devices, which we'll hear a lot more about in the next chapter. These allow the whole picture to be projected onto a rectangular array of sensors, each of which records the brightness of its own sample—picture element, or pixel—of the image.

In all cases, the brightness of the pixel is then converted into a discrete numeric value, called a quantization level. These values, as numbers, can then be fed directly to the computer or stored on tape for later processing. Depending on the way in which the photo was digitized, the information can be displayed on a CRT for direct viewing (or rephotographing if the picture is to be included in a magazine article or book), or used to expose a new photographic plate directly, without a TV-like image being formed first.

Once an image has been digitized and transmitted to a computer, the data can be operated on by various algorithms (sets of mathematical rules or procedures) to enhance the picture. It is the design of these algorithms that is the hard part. If each pixel is transformed the same way, such as by adding a constant, the result will merely be a brighter or darker picture. Of course, this can be used to compensate for over- or underexposure. Alternatively, each pixel value can be multiplied by a constant. This increases the contrast of the picture, by broadening the range of values. A picture with too much contrast can be reduced by dividing each value by a constant.

In much the same way, pixels with very similar values can be given strikingly different colors, as is done with false-color images. Thus, even when two areas of a photo are so similar in brightness as to be indistinguishable to the unaided eye, their difference can be made distinct by a computer.

The methods just discussed examine each pixel without reference to those surrounding it, but as useful as these methods are, they do nothing to unblur an out-of-focus picture, or one in which either the subject or the camera moved. To improve images of these kinds, the computer must select which pixels to enhance, which to suppress, and which to let alone.

Fourier-domain processing makes use of a mathematical procedure for analyzing signals. Fourier transforms are ways of manipulating and changing numbers, and they are applied to the ditigal data to detect how rapidly the brightness changes from one pixel to the next. The effect of this rather complex mathematical procedure is similar to what happens when one turns up the bass on the stereo. Only pixels with a certain characteristic, similar to sounds only in the bass range, are affected. In this way, an image that is out of focus can be sharpened, or the color qualities of a picture can be improved.

In many cases, however, it is necesary to first decide and define mathematically what caused the blurriness, and this is not always easy. Was the picture blurred because it was out of focus, because the object moved, because the lens was distorted, or because the air was turbulent?

Once the nature of the problem is defined, the proper Fourier-domain algorithms can be created to correct the image. It goes without saying that if this kind of mathematical operation were to be performed on a picture by anything other than a computer, the time involved would be tremendous. If you are processing not just one picture, but

the 1,500 or so pictures produced every week by the *Landsat* satellites, hand processing would be impossible.

THE ELECTRONIC "CAMERA"

In spite of all the things that can be done with traditional astronomical photography, they are not enough. Silver emulsion photographic plates can capture only so much light, require long exposure times, and are subject to errors in chemical preparation and processing. The data they contain, if they are to be handled by computers, must be converted into digital form. How much easier and more reliable it would be if, when we pointed the telescope at the sky, the computer could capture the image directly, without the frequently unreliable intermediate step of a photograph.

There are, in fact, a number of new methods of producing images from visible light that utilize microelectronics technology instead of film. Pictures made using these techniques can be enhanced and manipulated, just as photographs can, but far more easily and quickly, sometimes even on the spot.

One of these technologies involves a device called a vidicon, which is used to study the sun's corona, among other things. This machine is similar to a TV camera in some ways. The heart of the system is a chessboard-like array of photosensors on which the light from the sun (or another object) is focused. This array typically contains 10,000 photosensors in an area ¼-inch (.64-cm) square.

A Tiros weather satellite with two independent vidicon cameras on the underside

Each photosensor reacts to light, providing a highly sensitive and already digitized image. This image can be recorded on videotape, shown on a CRT, or sent directly to a computer, where it can then be enhanced in any number of ways. Or it can be printed as a photograph, before or after computer processing, for viewing by an astronomer or for publication in a book or magazine.

Many devices that help astronomers "see" the stars use photomultipliers, which are devices that can detect a single photon, or particle of light. When the photon strikes a photoelectric surface, it emits an electron that can be accelerated, generating a current, which in turn can be amplified to generate a measurable pulse of electricity. Again, a photograph can be produced, if desired.

Another type of device, called a charge-coupled device (CCD), works somewhat differently. This device makes use of the fact that when a photon hits a specially prepared silicon surface, it produces both an electron and a "hole" where the electron was. The electron is captured, and the "hole" is allowed to "disappear" by escaping into the material under the silicon.

Each electron is produced in a particular place on the CCD, and as the electrons are drawn off to one side, through a series of gates similar to those in a computer register, its original location is calculated. Some of these locations, called pixels, may contain no electrons at all, while others may contain quite a few. As they are drawn off, their number as well as their location is recorded. The advantage of this technique is that a CCD chip about 10 mm on a side, less than half an inch, can contain 250,000 pixels. Thus, even a very faint object can produce a strong image with incredibly high resolution and requiring only a very short exposure time.

The amount of data produced by such a device is staggering, which absolutely necessitates computer processing. Each pixel value is a computer word of many bits, not just a 1 or a 0. A single picture made by a CCD imager with 500 pixels on a side holds approximately the same amount of raw information as a 100,000-word book. And astronomers don't take just one picture a night but sometimes as many as a hundred. This is as much information as in some encyclopedias.

THE INVISIBLE EVIDENCE

Light is what we normally think of when we think of telescopes, and we can observe stars by visible light alone. Indeed, for a long time we were not even aware that other forms of radiation, except for infrared (heat, which is really just another form of light) existed. And although visible light is still perhaps the most common medium of astronomical observation, it remains a rather narrow band in the electromagnetic spectrum and tells only part of the story.

Astronomical objects emit energy in many forms. Radiation in every portion of the electromagnetic spectrum has its own story to tell about the object that emits it. Today, we know about these other portions of the spectrum—infrared, ultraviolet, X rays, and radio waves. But in order to make use of them, we need special devices that can detect them and convert what they detect into some form that we, as people, can understand.

Until recently, we could make no observations at all of certain radiations, such as X ray or ultraviolet, since our atmosphere largely blocks them out. It was not until the advent of high-altitude balloons, rockets, and satellites that we could put our instruments where they could do some good. Radio waves, on the other hand, are largely free to pass through the atmosphere, and radio frequencies have so far proved to be the most useful and most commonly used nonvisual portion of the spectrum.

RADIO SIGNALS FROM SPACE

A radio telescope is simply an antenna that detects a radio signal coming from a very small area of the sky. (This area is larger than that for visible light, since radio waves are longer than visible light waves.) Within each area, the signal has a certain strength, but it is impossible to capture this strength directly on film, as is done with visible light, so instead it is recorded as a number representing the value of that strength. The position of that area is also recorded, so that a kind of map can be drawn, with a number for each portion of the sky covered. The larger the number, the more intense the radio signal received.

Such maps are not very useful in themselves and would be tedious to compile and draw without the aid of a computer. But once the computer is called on to record each measurement and print out a map, it can also be used to convert those numbers into a pictorial image.

How does it do this? One way is by assigning a brightness level to each value, in much the same way as is done with some of the more sophisticated optical imaging devices. In this case, however, the brightness is a function of the strength of the radio wave detected at a particular frequency instead of a count of the number of photons received. Thus, the stronger the radio wave from a portion of the sky, the brighter the picture element, or pixel, that represents it.

Another way to map radio emissions from the stars is to draw lines connecting areas of equal intensity. This produces a picture like a weather map, or a topographic map; this kind of map can be more easily understood than a picture if the range of different values is large, say from −1,000 to +1,000, or the difference between different values is small, such as 6.75, 6.76, 6.77 etc. Indeed, the data, once stored in the computer, can be manipulated in many other ways than these two, as we shall see later.

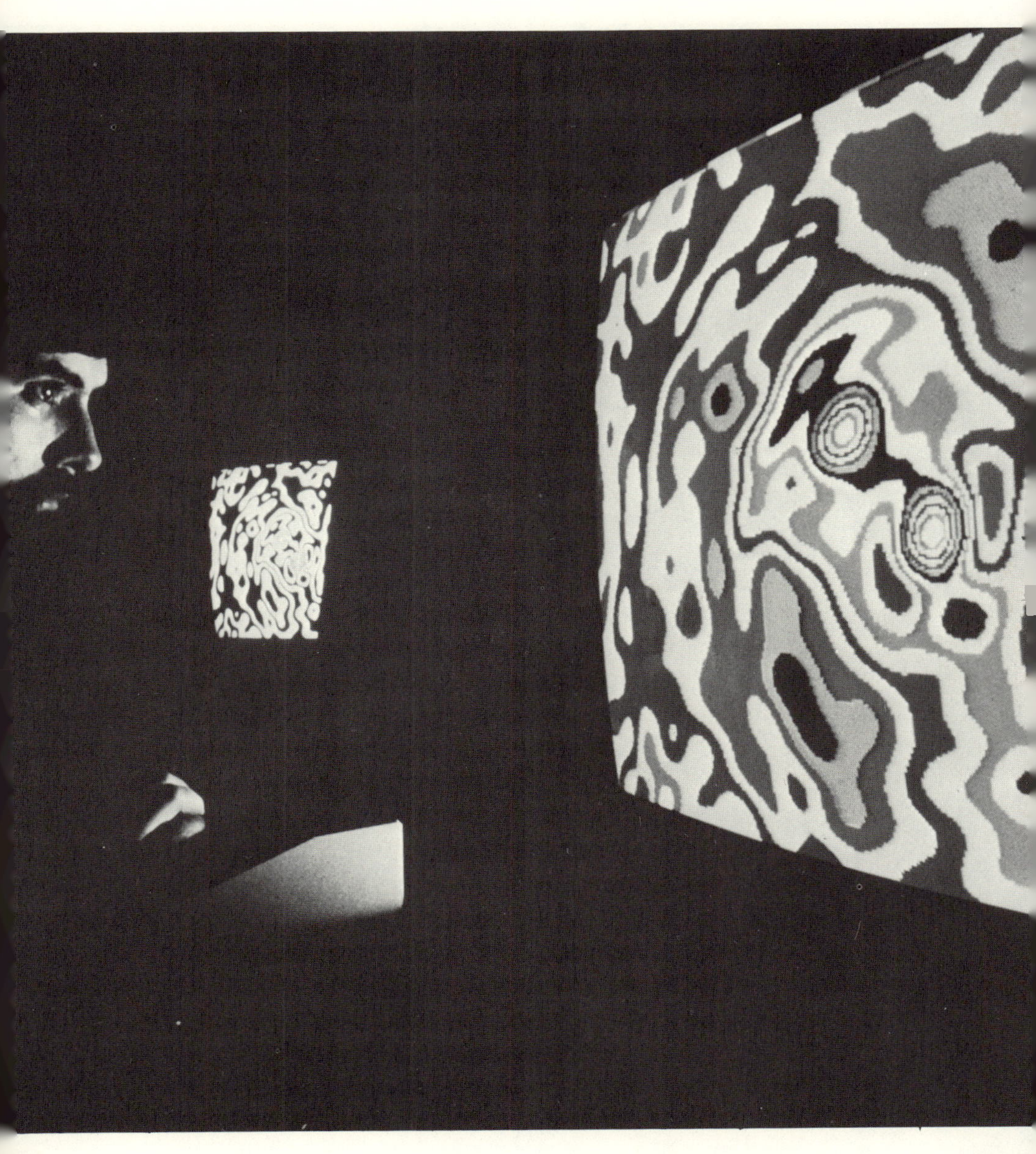

A computer-generated radio map of signals detected by the VLA

ARTHUR HILL

X-RAY DETECTION

Another form of radiation that is beginning to be studied closely is X rays. Because of our atmosphere, X-ray detectors must be carried high into the air by planes or balloons, or by rockets and satellites. *Copernicus*, the Orbiting Astronomical Observatory launched in 1972, is one of these.

The orbiting *Einstein Observatory* also detects and processes X rays. Sophisticated computer-imaging techniques are then used to translate the signals into pictures. These and other X-ray observatories and telescopes can tell us a lot about astronomical objects that are otherwise invisible.

X-ray telescopes do not expose a photographic emulsion, as is the case with X rays used in hospitals and dental offices. Instead, the radiation is measured by a special detector, and the intensity within a minute area is recorded.

One of the difficulties with X-ray observation involves focusing the incoming rays. With light, lenses and mirrors do the job. With radio waves, the antenna dish serves as a kind of mirror. But X rays pass through most substances and are not easily reflected or bent. It has been found, however, that X rays striking certain materials at a very

An artist's concept of NASA's Einstein Observatory, *which was launched in 1978. The orbiting observatory has an X-ray telescope that picks up X- and gamma radiation, which can be turned into photographs.*

shallow angle, merely grazing the surface, as it were, can be deflected slightly. This is the principle behind the grazing-incidence X-ray telescope.

Copernicus has three such telescopes on board, each sensitive to a different range of energies. The collected and focused rays are directed to detectors that convert the signal strength into digital data, which is then transmitted to ground stations. There it is processed and reconstructed, giving us a low-resolution image of what a portion of sky might look like if we could see at X-ray wavelengths.

To make it easier for us to understand these reconstructed pictures, they are displayed on a CRT in false colors; that is, the X rays of shorter wavelengths are arbitrarily assigned colors toward the blue end of the visible spectrum, and those that are longer are represented as being red. A scale from black to white could as easily be used, but colors are more meaningful to humans.

MORE IMAGING OF DATA

As data from observational equipment is nowadays more frequently being recorded in the form of electronic pulses, instead of physical and chemical changes in a photographic emulsion, the imaging ability of computers—that is, the creation of pictures from nonpictorial data—becomes ever more important and useful.

The quality of the pictures so produced can be changed and adjusted, under computer control, as the data is being received. No photographic intermediate step is needed (though it is sometimes included), only a CRT on which the visual image can be displayed. Not only can the computer enhance resolution, or contrast, or eliminate certain intensities or wavelengths, it can also change the shape of the image, invert it, or change the perspective.

If properly programmed, the computer can manipulate

the incoming data to show only specific frequencies in false colors. Or it can combine two or more pictures, such as the current observation superimposed on a reference image, or an image taken in the infrared wavelength superimposed on one in the ultraviolet wavelength.

Sometimes, two different processes produce light (or other forms of radiation) that are very similar in wavelength. Our eyes, if we could see this light, might not be able to distinguish between these two frequencies, but the computer can assign each of them a distinct false color. At other times, light of one form is so bright that light in another color is obscured. Thus, by eliminating all visible light and showing only the infrared, we see a different picture. We can also look at the light coming from a rotating body, whether the body itself is producing that light or only reflecting it. The light coming from the side that is moving away from us will be slightly redder than the light coming from the side that is moving toward us. By comparing the difference in color, we can determine its rate of rotation. In this case, no picture need be produced at all.

Manipulations of this and other kinds become even more important when the portions of the spectrum we wish to study are outside the narrow range of visible light. Given the right equipment, any energy frequency can be detected and converted into a picture. This is possible even without a computer, in many cases, but with a computer's assistance the data can yield much more information.

FILTERING THE INCOMING DATA

Sometimes it is impossible to make a "clean" observation. That is, other stuff—"noise"—comes in with what we want. Our detecting equipment has limitations as to how much of this noise it can filter out, but computer analysis of the data, and manipulations of that data similar to those used to enhance photographs, can clean up the observation considerably. More important, only the computer can do this fast enough. When data is coming in at millions of bits a second, hand calculation is just not possible.

Sometimes, the observation is not bothered by noise so much as by the fact that too much data is coming in. Radiation is produced along the whole spectrum, sometimes from similar processes, sometimes from different ones. We could narrow down our observation to view just one process at a time, but sometimes we don't have an opportunity to make a similar observation later. Thus, we collect as much as we can and let the computer sort out the different things going on. The computer "filters" the data, so that each process can be examined separately.

REAL-TIME ADJUSTMENTS

Telescopes see what is in front of them, and a large part of their control lies in making sure that they are pointed the right way. But there is more to it than that. Sensors and instruments collect incoming radiation, whether it be light or radio or whatever, and convert it into digital data. Computers then use the information inherent in the strength or type of signal to control the aperture and "exposure" time. Although these things can be set beforehand, the computer allows on-the-spot, moment-to-moment adjustments, using its ability to make decisions based on current conditions.

The computer does not wait for humans to tell it what to do. Even as the data is coming in, the computer can sample that data and use it to improve reception, enhance certain features, eliminate blurs, and so on. It can accommodate for jerks in the telescope, dust or moisture in the air, and other problems even before putting the image on a screen. Such real-time adjustments and corrections are just not possible with traditional photography.

PREVENTING OVEREXPOSURE

Given a sensitive enough detector, it may take only a fraction of a second to make a given observation. If that is true, observation times that are too long will result in a meaningless "overexposure." Mechanical timers have limitations in the duration and accuracy of the times they can measure. Computers, on the other hand, can measure down to thousandths or millionths of a second, and with great precision.

This can be of considerable importance when observing something like a pulsar, which flickers hundreds or thousands of times a second. The computer can measure this flickering precisely. It can compare a series of many sequential observations, and even if it never quite catches the beat of the pulsar, it can measure the sometimes tiny changes in brightness in such a way that its period, or the time cycle between its brightest moments, can be accurately determined.

* * * * * * * * *

The sum of what we learned about the nature of the universe, before the advent of the computer, is quite impressive, but it is nowhere near what we are learning now and will learn in the future, with the computer's help. Before, we were limited by the time it took to get a good photograph and analyze each image. Now, by using electronic instead of chemical techniques, we can map practically the whole galaxy from earth, even parts that are "invisible" to us because of intervening gas, obscuring energy, and so on.

CHAPTER 3

SPECIAL TELESCOPES

THE SPACE TELESCOPE

Human control of telescopes in space is virtually impossible. Every detail of aiming, aperture, exposure time, recording, translation, and transmission must be seen to by onboard equipment. Remote control from earth is possible for operations that require little detail, but most astronomical observation requires extreme precision. The onboard computer is loaded with a program from earth and reprogrammed as necessary, but the computer itself does the work.

The importance of computer control of all aspects of orbital observation will become obvious with the launch of the Space Telescope in 1985 or 1986. This observatory will be packed with highly advanced electronic instrumentation for the detection and processing of visible light. The onboard computer will handle all the details, according to

An artist's concept of the Space Telescope, due to be launched in 1985 or 1986

The 94-inch-diameter primary mirror for the Space Telescope (center) *in a huge thermal-vacuum chamber at the Perkin-Elmer Corp., where fine polishing was done under computer control*

its programming, including aiming the telescope, maintaining its position, selecting the kind of sensing device or camera to be used, controlling the duration of the observation, converting light signals into digital data, and transmitting the data back to the earth.

About the only remote-control aspect of the Space Telescope will be that its computer can be reprogrammed from the ground. This means that as we learn more about the objects the telescope is looking at, and about how to manipulate the data it produces, we can upgrade the telescope's performance. It becomes a very flexible, as well as very powerful, tool, not constrained to one mode of operation.

The diameter of the Space Telescope's main mirror is only 8 feet (2.1 m), in order that it be small and light enough to be launched by the space shuttle. This is not as big as many telescopes on earth, but the advantage is that, out in space, there is no atmosphere to degrade the image.

In order for the instrument to be fully efficient, the surface of the mirror has to be as smooth as possible and conform to the correct shape as closely as possible. On earth, such perfection is not worth the effort, since atmospheric distortion is greater than that caused by the mirror's imperfections.

To achieve as perfect a surface as possible, engineers at the Perkin-Elmer Corporation, which built the mirror, used computer controls to polish it. Since it was made on earth, under the influence of gravity, they had to devise a method of predicting how the mirror would deform under conditions of weightlessness, so that their polishing could take this into account.

After each of the twenty-four cycles of the final, month-long polishing process, a special interferometer reflected light off the surface. From this, computer imaging produced pictures of the surface that showed deviations from the desired final shape. When the polishing was done, the maximum deviation was less than one millionth of an inch over most of the surface. If the United States were that flat, hills or valleys would be less than 2½ inches (6.25 cm) high or deep.

The glass disk was then put in a vacuum chamber containing slugs of aluminum, which were bombarded with electrons from an electron beam gun. This caused the metal to boil off in a cloud that then coated the mirror, providing its reflective surface.

The Space Telescope will use CCDs for part of its imaging system. One of these is the wide-field/planetary camera. At one level of resolution, this camera can take a picture of a large portion of the sky, the "wide-field" operation. At a higher resolution, it narrows down to a smaller portion of the sky, but with a magnification that should enable it to detect planets around nearby stars—the "planetary" operation.

The other part of the telescope's imaging system uses an image-intensifying device similar to the light-sensitive CRT in a TV camera or a photomultiplier. Using this technique, the faint-object camera can count individual photons.

The Space Telescope, besides its optical devices, carries two spectrographs: faint-object and high-resolution. It also carries a high-speed photometer which, though it does not produce any pictures itself, can detect events of 10 microseconds duration.

This is all pretty high-tech stuff, and some of the equipment costs millions of dollars. But even amateurs are getting in on this new form of imaging.

The Independent Space Research Group (ISRG), based at Rensselaer Polytechnic Institute in Troy, New York, proposes to send up their own Amateur Space Telescope on the space shuttle. It would be smaller than NASA's Space Telescope but would use a highly advanced digital imaging system, like those used at the principal observatories on earth. Light entering the telescope would be converted to digital code and stored in a computer. Then on command, the code would be sent back to earth, where another computer would reconstitute the image. This is a student-designed and student-operated program. The ISRG has plans for other telescopes as well, using ultraviolet light and radio frequencies.

THE MULTIPLE MIRROR TELESCOPE

The larger the mirror of a telescope, the farther it can see into space. But large mirrors are difficult and expensive to make, and their supports and sheltering domes are massive. There is an alternative, in which a number of smaller mirrors are made to work in unison, all aimed at precisely the same point. Such a telescope is the Multiple Mirror Telescope, or MMT, at the Smithsonian's Mount Hopkins Observatory in Arizona.

The MMT has six Cassegrain telescopes arranged in a ring. Each telescope has a 72-inch (183-cm) mirror. The light-gathering power is equivalent to an aperture of 176 inches (447 cm), since all the mirrors work together. Light, reflecting off the large mirrors, is brought to a focus by smaller, movable mirrors under the control of a computer. To help in this, a laser shines a spot of light on the mirrors, and the computer moves the smaller mirrors until all six spots are in the same place. If this were not done, the different large and small mirrors would produce a fuzzy image, worse than one large mirror alone.

This makes the MMT, in effect, the world's third largest telescope. The only two with greater light-gathering power are the 200-inch (508-cm) Hale Telescope at Mount Palomar, and the 236-inch (600-cm) Soviet telescope in the Caucasus. Both of these are single-mirror telescopes and so do not need to be finely focused by computers.

Astronomers had been acquainted with the concept of multiple-mirror telescopes long before the MMT was built.

The principles that enable such a telescope to have an effective aperture larger than the individual mirrors are similar to the ones used in interferometry, which we'll discuss in some detail soon. But astronomers could not form and hold steady a single image from several mirrors before computer control made it possible. Now, using lasers to form guide images, automatic electronic sensing, computer control systems, and other forms of sophisticated electronic technology, that single image is possible.

In the MMT, lasers keep the telescopes in line by simulating a star that the guidance computer can detect and line up with. Thus, the telescope can be aimed at an object that is otherwise invisible. Each of the individual Cassegrain telescopes can be adjusted by the computers independently, depending on the depth of field desired. Computers also allow tracking on a two-axis mount, to compensate for the motion of the earth and, if necessary, any movement of the object of investigation. Indeed, the entire building that shelters the telescope—and the computers and administrative offices—rotates under computer control.

Given the power and flexibility of the MMT, its cost was quite low, about $8 million. A conventional mirror with conventional controls of comparable size would have cost about three times as much.

Theoretically, there's almost no limit to the effective size of a multiple-mirror telescope. Single mirrors are constrained by mass, the effects of gravity, changes in temperature, and the perfection of the reflecting surface. But small mirrors are relatively easy to make, and to make well. They are far cheaper than large mirrors and are less subject to distortion due to their weight. More than six mirrors could be made to work together in a similar way, providing even greater light-gathering power than the MMT presently does.

And with computer control of the telescope, the mounts, and the dome, astronomers no longer have to sit

The Multiple Mirror Telescope sits atop the summit of Mt. Hopkins in Arizona's Santa Rita Mountains. The entire building—including offices, control room, and laboratories—turns with the telescope as it tracks the stars.

Two dishes of the Very Large Array radio telescope near Socorro, New Mexico

in the cold and dark for hours on end, staring through an eyepiece and delicately turning thumbscrews. They can give the computer a program, set the telescope up, and come back when the job is done.

The MMT does not use film to record the image. Instead, the light passing through the multiple mirrors is directly recorded as digital data, which is then interpreted and displayed by a computer. The computer does the job by using self-scanning diode arrays, some with 10,000, others with 100,000, individual light-sensitive elements on a silicon wafer the size of a fingernail. Needless to say, the resolution of these images is higher than that of a television image or the finest newspaper picture, or even of many photographs.

The MMT is equipped to do visible light, radio wave, X-ray, and infrared observations, among others. It also has a special red-shift machine that uses a CCD-imaging system to measure the apparent speed of distant objects moving away from us.

THE VERY LARGE ARRAY TELESCOPE

The MMT was inspired by radio telescopes with more than one dish. Like the MMT, such radio-telescope arrays have an effective aperture much greater than is possible with a single dish. The biggest and most famous of these is the Very Large Array, or VLA, near Socorro, New Mexico.

There are twenty-seven dishes in the array, each 82 feet (25 m) across and weighing 212 tons. They are arranged in the form of a huge Y, with an effective aperture of 21 miles (33.6 km). In order to achieve this artificial aperture, each dish must be pointed at precisely the same place in the sky and must follow that point as the world turns.

Radio signals from astronomical sources that reach the twenty-seven dishes are directed by a subreflector to one of four sets of receivers, amplifiers, and signal processors, each of which is designed to do a special task. These produce a stream of numbers, each representing a signal strength at a certain wavelength within an area of sky just a few tenths of a second of arc in diameter.

The signals are then transmitted to the central control building for computer processing. A waveguide is used instead of cable, in order to ensure the integrity of the data. A waveguide is a hollow metal tube, inside which the signals travel, like light along a fiber of glass.

As soon as the signals are received at the computer building, they are processed by a method called interferometry. Images or signals received by antennas at different locations are not quite identical. When combined together, they interfere with each other, the way waves from two pebbles dropped into water interfere with each other. It is not as simple as that, of course, but this interference phenomenon, after it has been properly processed and interpreted, can convey considerable information to the observer.

There are four minicomputers used at the VLA for all kinds of purposes. These have been given fanciful names, and each is assigned a range of tasks. The computers not only control the dishes but also turn the data into images or produce other forms of output if desired. The one named Boss oversees two other minicomputers—Cora, which collects the signals from the antennas, and Corbin, which correlates the data. The fourth minicomputer, Monty, monitors all performances.

VERY LONG BASELINE INTERFEROMETRY

In the VLA, all the radio dishes are connected directly to a central computer system. The distance between the central processing station and each dish is known precisely. This is necessary for the proper evaluation of interferometric information. But radio telescopes used for interferometry do not have to be physically linked.

Very Long Baseline Interferometry, or VLBI, uses radio telescopes that can be half a world apart. Networks of telescopes up to 6,000 miles (9,600 km) apart have actually been used. If these telescopes could be physically connected, the precise length of the connecting wires would be known, and the signals from them could be easily coordinated. But such a connection is not possible; the six or seven antennas stretch from Europe to North America, and from South America to Australia, with oceans in between.

Instead, the time of each signal's arrival at the scope is calibrated by very precise atomic clocks. That timing information, along with the observational data, is recorded on magnetic tape and then sent to computers that will process the images at a later time. The clocks guarantee that the recordings are synchronized to within a tenth of a microsecond, that is, one ten-millionth of a second. Any greater error would garble the picture and make it meaningless.

Images produced by VLBI require anywhere from one to ten trillion bits of raw data (10^{12} to 10^{13} bits). With VLBI, astronomers can get angular resolutions of 300-billionths of a degree, or better. But the amount of data is so large that twelve hours of observing time can require up to 120 hours of computing time. The computer processing is done at the National Radio Astronomy Observatory in Greenbank, West Virginia; at Caltech in Pasadena; at the Haystack Observatory in Westford, Massachusetts; or at the Max Planck Institute in West Germany.

The technique of interferometry is not restricted to images or signals received at different locations. Those received at different times can also be used. A single radio telescope can observe a star in the morning and then again in the evening. The time lapse must be precisely measured, of course, but the effect is the same as if two different telescopes in different places were used, since the earth's rotation and orbit move the antenna through space.

ARECIBO, SOLAR MAX, AND IUE

The largest and most famous of the single-dish radio telescopes is that at Arecibo in Puerto Rico. This dish, built in a natural valley, is 305 meters, or over a thousand feet, across. It is used for all sorts of projects, including the Search for Extra-Terrestrial Intelligence, or SETI. Needless to say, the antenna is controlled by computers. And computers record the millions of bits of data and store them for later study. Arecibo, under computer guidance, sent a binary signal to the globular cluster M13 in 1974, a message that was not really intended to be received but that served mainly as a symbol of our technological capacity to make interstellar radio contact with other civilizations.

The Solar Maximum Satellite was put into orbit to take pictures of the sun and its corona. It does not use photography, though its sensors are designed to detect visible light. It converts the image its sensors detect directly to digital data that is then transmitted to the ground, where a computer reconstructs the image and prepares pictures and analyses.

The International Ultraviolet Explorer (IUE) is a satellite in nearly geosynchronous orbit, a joint enterprise of NASA, the British Science and Engineering Research Council,

The single-dish radio telescope, the largest in the world, at Arecibo, Puerto Rico

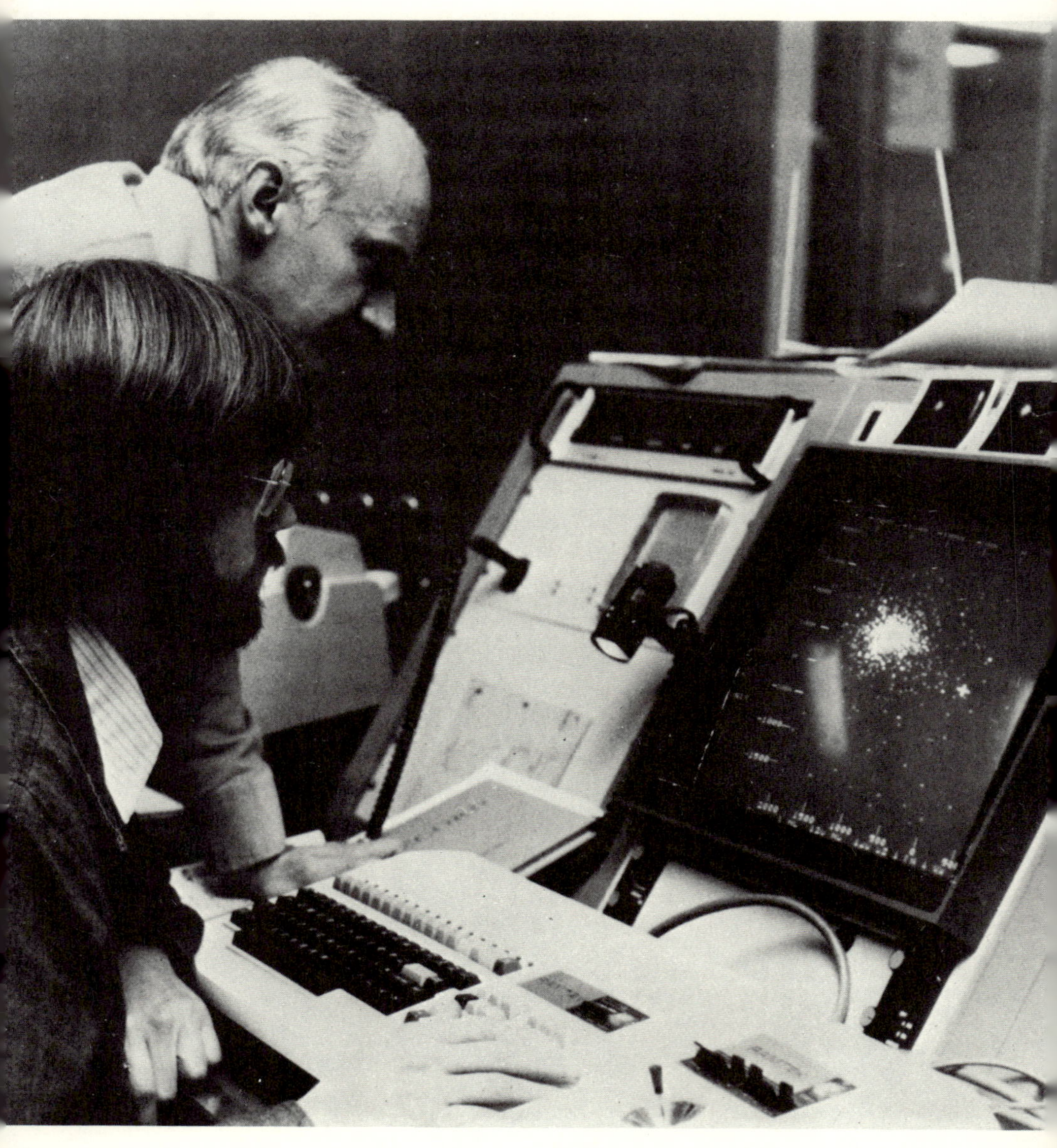

NASA workers watch a viewing screen on which appears an image of the first star observed by the International Ultraviolet Explorer (IUE).

and the European Space Agency. During the first two years since its launch in January 1978, it has taken more than 30,000 photographs. These are not really pictures, however, but special UV spectral images.

The IUE's telescope has an 18-inch (46-cm) aperture that feeds light to either of two spectrographs, instruments that separate and organize the radiation from a UV source into spectral bands that correspond to wavelength. Then, a UV-sensitive TV camera photographs the bands for relay to earth by digital signals. The image is then reconstructed, colored, and enhanced by a computer. The images are like the familiar picture of the visible spectrum but are manifest only in UV wavelengths. The color coding is designed to enhance special features. The resolution or wavelength range can also be changed, but only after the data is received. The images are not what one would see but are graphic portrayals of energy.

CHAPTER 4

ROCKETS INTO SPACE

Though telescopy is such a huge part of space science, what many of us think of when we hear "space science" mentioned is rocketry and spacecraft, which are used to get our instruments, including telescopes and cameras, up out of the murky sea of the atmosphere and closer to the objects of our interest. Computers have played a large part in the process of extending our vision and other senses into the depths of space.

Computers are used extensively in both the designing and manufacturing of the mechanical parts, electronic equipment, and other computers and microcircuits for rockets, satellites, and communications systems. More than that, they also help in figuring out how all the parts can interrelate with each other, how the equipment should be powered, and how it should be used. Since another book in this series covers the subject of computer-aided design and manufacture more thoroughly, we will limit our discussion here to only those aspects that relate directly to space science.

This is appropriate because the design problems in modern space science are sometimes far more complex than in other areas of industry and are frequently beyond the ability of human minds to solve unaided, or within a reasonable period of time. Computers may be used to design an automobile part, but a properly trained person could do the same without computer assistance, though it might take longer. But in designing the shape of the space shuttle, so many factors apply, both known and unknown, that only computerized simulations can provide the quality of design necessary for success.

Also, the parts manufactured for lunar landers, communications satellites, and space capsules are frequently one of a kind. This imposes further difficulties not only on the design but also on the manufacturing process. Many of these physical components must have tolerances impossible to achieve by hand, and only highly automated and computer-controlled machinery can produce them, as was the case with the mirror for the Space Telescope.

In order to build a spacecraft, every step of the design and manufacturing process must first be specified, indicating what resources in materials and labor will be required, how long each step will take, which steps are dependent on others, and how much each part of the project will cost. Engineers and scientists decide what they want done, and the computer helps them determine whether their objectives are feasible, or even possible, given the available resources, time, money, and the current state of knowledge.

Special techniques of scheduling have been developed which, for small jobs, can be performed with pencil and paper, but for something as complex as an *Apollo* mission or a *Voyager* spacecraft, a computer's assistance is required. And, of course, computers help in keeping track of everything as it occurs, including paychecks and financing. Thus, even business computing has a part to play in our observation and exploration of space.

COMPUTER-AIDED DESIGN

It is, of course, possible to build a rocket without a computer entering into the process at all. Many small research rockets that probe the upper atmosphere can be built by a machine shop using well-known and well-understood principles. Amateurs can build their own model rockets, from kits or from scratch. And they will all (theoretically speaking at least) work.

But the importance of good design becomes apparent when, using commercially available rocket-kit parts, one builds a model rocket of original design. The diameter of the body, its length, and the location, size, and shape of the fins are all critical, and two rockets of the same mass but having different shapes and proportions will fly differently. The difference in performance of two such model rockets is interesting, but in a mission costing millions of dollars and perhaps with astronauts on board, such differences are critical.

Our largest launch vehicles do not look all that special—just cylinders with engines on the bottom and payloads on the top. But their performance under the stress of takeoff and during their passage through the atmosphere must be thoroughly understood in order to avoid costly mistakes. The functioning of the engines, the delivery of fuel, and the successful separation of the stages must all be carefully planned for. And although launch vehicles may look clunky, satellites and spacecraft intended to return to earth intact must be very carefully designed.

Thrust, weightlessness, vibration, friction, heat, aerodynamics, center of mass, shape, and other parameters all combine together in a highly complex and interrelated fashion. These factors must all be taken into account, as they are and as they change with time, distance, and other conditions. The success of a mission, the survival of the craft, and the safety of the people on earth on whom a failed craft might fall all depend on the successful design of all these elements. Just the sheer number of these elements, and how they interrelate, demands the assistance of a computer, which not only keeps track of everything but also performs real-time calculations as well. The results of these calculations are sometimes implemented automatically, by the computer itself, and are sometimes presented to humans for their decisions.

USING COMPUTER SIMULATIONS

Computer simulations were used extensively in testing the various designs of the space shuttle as well as its launching engines and external fuel tanks. Although certain tests could be performed with physical models, these tests were controlled by and recorded by computers, which then analyzed the results.

Similar testing and simulation is done for all large or experimental aircraft these days. Even models cannot reproduce exactly what will happen, since air viscosity, or its relative "thickness," is different for a small model than a large one. Think about a movie in which the ship at sea during a storm is represented by a model. The waves splashing against the model don't look like those you would see buffeting a full-sized ship. We can reduce the scale of the ship but cannot regulate with precision the power of the waves. It's the same with air. Of course, the only real test is to send the rocket up and then, with computers monitoring everything possible, record the data and analyze it.

Always before a launch, the proposed flightpath and orbit of a satellite are simulated. Given the mass and velocity of a rocket, and the known effects of the earth's atmosphere and gravity, orbits can be computed to

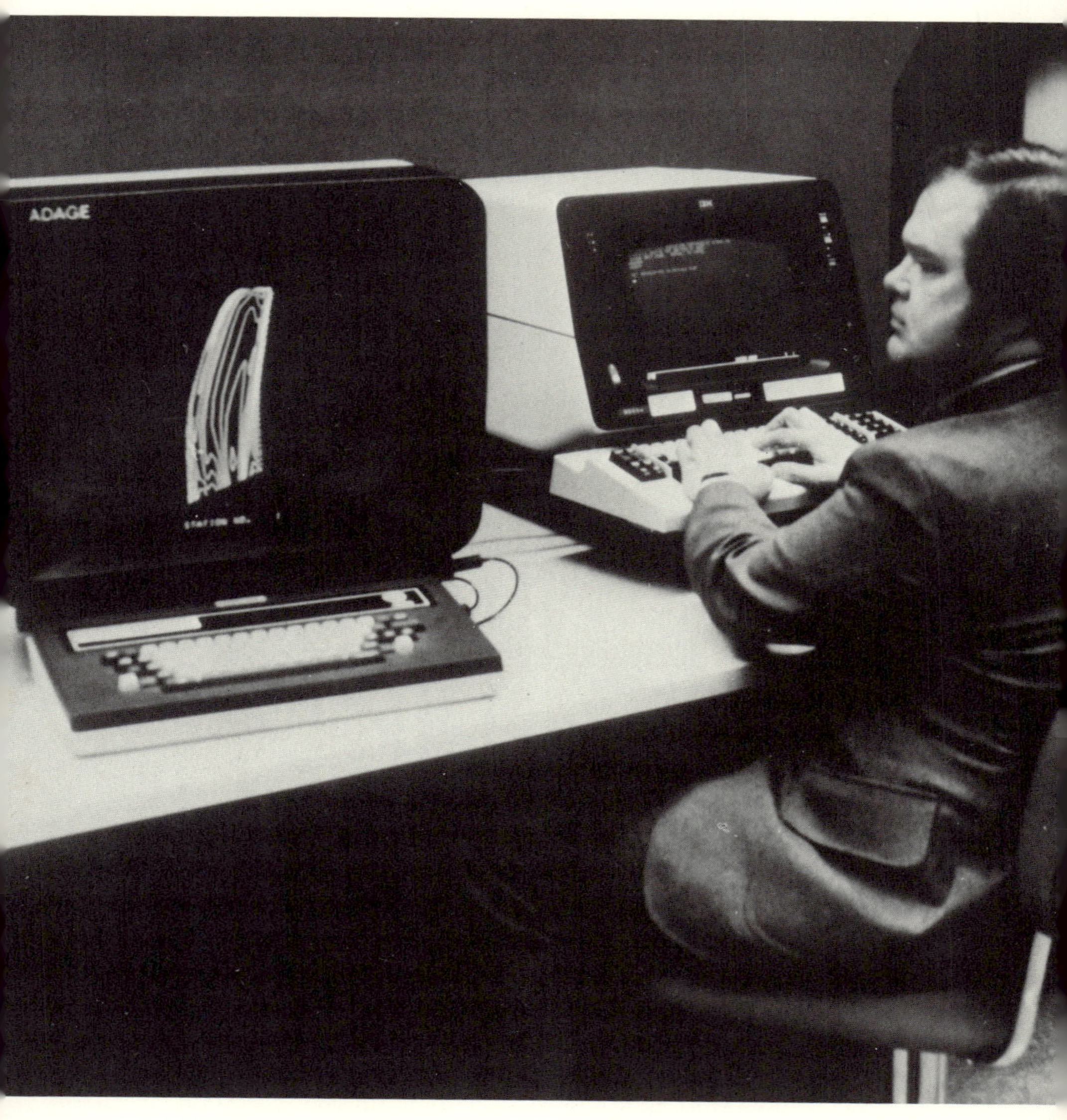

Computer graphics that simulate the effects of varying air speeds are used in designing rockets and other high-speed aircraft.

achieve the greatest fuel efficiency, easiest insertion into orbit, greatest security, and so on. Real launches must be made, to test the theories, and then the theories must be modified to create mathematical models, to simulate dozens of hypothetical launches with varying conditions. In sending a rocket to another planet or the moon, the exact time of day, angle of flight, departure from orbit, course corrections, and so on must all be calculated in advance.

Every piece of equipment on board a shuttle, satellite, or probe must be tested. Sometimes, special materials are required. Other times, the design of a piece of equipment can be improved by redesigning it, but that requires tests first. In many cases, theories of statistics and probability must be used to determine the meaning of a sequence of perhaps hundreds or thousands of tests. This can be done by hand, but it is impractical. Instead, a computer is used.

The same is true for all experimental systems, such as life detectors, soil samplers, energy detectors, and so on. The systems must be tested on the ground before the expense of a launch. Again, of course, physical tests can be performed in many cases, but data from a series of tests, especially tests that involve several pieces of equipment, must be collated and compared, to see how the system works as a whole, not just as individual parts.

Ultimately, of course, comes the final test—launch.

THE LAUNCH

Shuttles and other spacecraft must not only be able to survive a return trip through the atmosphere, they and all their equipment and passengers must also be able to withstand the vibration and thrust of takeoff and the conditions of weightlessness in orbit. A further problem at takeoff is that as the rocket flies, its mass changes as it burns fuel.

Its shape changes as it drops off stages. Accounting for these changes requires computer analysis and control.

Computers monitor the rocket's functions at all times during a launch, ready to make corrections if necessary as well as record the overall performance of the vehicle. As the craft moves through space, computers control ground-based antennas to keep them aimed at the vehicle.

The actual firing, feeding of fuel, and rate of thrust are controlled in a variety of ways. Some function in a clockwork fashion, that is, nothing can change once a process has started. In other cases, a human controller can throw switches or turn dials and thus affect how the rocket works. This is just remote control. The position and speed and other functions of the vehicle are monitored in various ways, and if something is amiss, a signal is sent to the rocket to correct the error. Even here, ground-based computers are used, since much of the remote-control work is handled by computers as opposed to humans. This is also true in large part for many of the operations of a satellite in orbit.

Many of the functions of a rocket, however, are controlled by onboard computers, designed for just that purpose. With an expensive rocket we can't take the chance of there being interference with a radio signal, and so wherever possible or necessary, a program stored in a tiny (relatively speaking) computer on the vehicle itself is used to monitor fuel, thrust, and attitude, as well as the functioning of the electronic components and communications systems. These computers can be overridden by instructions from the spaceflight center, if necessary, or they can be reprogrammed at a later time.

Solid-fuel rockets are not easy to control. The fuel, once ignited, just burns. If any adjustments are to be made, they must be included in the original design of the fuel charge, rather than saved until the rocket takes off. The rate of fuel flow can be adjusted with liquid-fuel rockets, however, and the direction of thrust changed. This is especially true of the smaller rockets and jets that control the vehicle's attitude and direction.

A second stage must be timed perfectly. Theoretically, it would be possible just to set a timer, or install a device to detect when fuel in the first stage has run out, to start the second stage burning. In practice, this just doesn't work. The first stage must be blown free, and this can impart unwanted motion to the rest of the rocket. Also, the attitude of the rocket and its altitude and velocity can affect whether or not the second-stage engines will be able to do what they are supposed to do. The second stage fires only when conditions are right. Of course, every effort is made before the launch to make sure that those conditions occur as a matter of course. Still, onboard computers are provided with programs that can make adjustments, to take into account anything that has changed the performance of the vehicle up to the moment of separation and firing.

After a while, the major portion of the fuel is burned, and the rocket coasts into orbit. Even if it is going on to the moon or to other planets, one orbit is usually flown to check that everything is working right and to make sure that the precise direction of thrust is applied. Maneuvering into orbit is a tricky business, requiring more than one thrust of the engines. The computers on board and on the ground help control this, to achieve the orbit desired.

IN ORBIT

The orbit of a satellite is adjusted and corrected after observations by ground-based tracking stations. These stations make precise measurements of the location and speed of the satellite, and the data is analyzed by computers on the ground, which can then transmit instructions to the much smaller onboard computers.

Ground tracking of satellites for orbit correction is one possible weak link. As a craft orbits the earth, it passes out of range of the original antennas, and control and message signals going either way must be redirected to other antennas located in other parts of the globe. The time delay here must be accounted for and managed by computers.

Atmospheric conditions can interfere with transmissions, as can certain solar effects. Messages and signals are affected by the weather, which can produce noise or blot out whole portions of the signal. Computers can analyze these fuzzy or broken messages, using communications theory (the mathematics that describes how information is transmitted and affected by transmission), and sometimes reconstruct all or part of the missing information, but anything that cuts the remote-control link jeopardizes the operation.

CHAPTER 5

THE SPACE SHUTTLE

Our most ambitious near-earth operation has been the space shuttle, which depended on computers for all phases of its development. One of the special problems with the shuttle was providing a comfortable living and working environment for those who travel in it. With human lives, millions of dollars, and perhaps the future of the whole space program at stake, few chances could be taken. The people who pilot the shuttle, and who control and direct its experiments and other operations, must be not only well trained but also capable of performing at all times, and this required a thoroughly designed life-support system.

Because of space limitations and the lack of gravity, providing this high-quality life support was not easy. Even such a homely device as the toilet had to be specially designed. On the shuttle, if the toilet backs up, the consequences can be catastrophic, and you can't just call a plumber.

The foods astronauts eat, the air they breathe, the water they drink, and the systems that prepare and deliver all these things are tested thoroughly, and computer analyses are made. Even the shapes of the switches and the locations of data screens and handholds were carefully considered. Real-life tests were, of course, indispensable, but proper preparation and planning were also necessary, and computer-aided design and manufacturing played a large role.

The space shuttle's mission simulator at NASA's Johnson Space Center outside Houston is a $100-million device. It uses computers to simulate the onboard computer and instrument responses, the motions of the space shuttle, and the visual scenes of flying in it. Everything the astronauts will experience is simulated precisely, even takeoff acceleration.

THE FLYING BRICK

The space shuttle has been called a flying brick, and the metaphor is an apt one. A brick has no aerodynamic characteristics to speak of, and the shuttle, in its overall shape and mass, resembles a brick far more than it does a sleek, modern airplane.

But the shuttle *is* aerodynamic. It can fly, even glide in a manner of speaking. And that is because its shape has been so carefully designed that it responds even to the effects of traveling through the atmosphere. In order to achieve this aerodynamic capability, scientists and engineers had to make extensive use of computer models, as well as of physical models.

Full-scale physical models of each design variation or change would have taken far too much time to build and would have been far too costly. Instead, by use of computer modeling and by incorporating into the program the laws of aerodynamics and their effects on the shape and

The Space Shuttle mission simulator

mass of the shuttle, many inappropriate designs could be eliminated in advance, leaving only those most likely to succeed to be made into physical models and tested in a wind tunnel (and later in a full-scale test).

Wind-tunnel tests are needed, since the flow of air around a moving body is so complex that a computer model cannot yet simulate it completely. Some engineers estimate that computers would have to be about 10,000 times as fast as they are now in order to handle all the variables involved in real time.

With a physical wind tunnel, you can see what is happening, though of course if the model is not full size and full mass, it will not behave exactly as the real object would. With a computer model, on the other hand, the simulation is only as good as the data from which it is derived, and since fluid dynamics is one of the most complex areas of physics, this data is not always well understood. Some events take place in fractions of a second but would need many minutes or hours of computer time to simulate.

Still, computer simulations are helpful. The mathematics describing the flow of air around an aircraft, or through a jet engine, are incredibly complex. Called Navier-Stokes equations, these have been known for over a century but are so complicated that, until recently, they were used only in special situations. Working them out by hand, even with the aid of calculators, took hours or days or weeks, and every change in design required a reworking of the equations.

Now, with modern high-speed digital computers, these equations can be much more easily solved. Even using an antiquated computer such as the Illiac IV, built in 1970, engineers at the Ames Research Center, in California, can study these equations in an effort to design a new kind of jet engine, called a scram-jet, which might eventually be used instead of today's launch boosters to put the shuttle into orbit.

ONBOARD COMPUTERS

Because the shuttle is aerodynamically somewhat unstable, five onboard computers are needed to control it during atmospheric flight. Each has an input-output processor, a central processing unit, and measures 7.5 by 10 by 19.5 inches (about 18 by 25 by 54 cm). These do their job during the first hour and a half of ascent and the last forty-five minutes of re-entry.

Four of the computers operate in concert as a redundant set, with the fifth as a backup in case of malfunction. Each time they complete a stage in their computations, the four computers synchronize with each other—about 300 to 400 times a second. They finish their synchronization within 50 to 100 microseconds (millionths of a second) of each other. A lag of four milliseconds (thousandths of a second) means that the delaying computer is assumed to be malfunctioning, and it is switched out. The ship can run with only three computers operating, but if a second fails, the flight is terminated and control is given to the fifth, special computer, which brings the ship back.

These computers are rather old-fashioned but very reliable. They were designed in 1972. Four hundred people worked for four years writing the software for the *Columbia*, producing 350,000 lines of code.

On board the shuttle are three CRTs. One provides digitally printed data on the spacecraft's systems; the other two provide trajectory plots, a kind of roadmap of where the craft is going. Needless to say, these are "intelligent" CRTs, computer-controlled not only to produce the images but also to interpret the data sent to them.

Computers control the shuttle's rotation of 120 degrees, to align the ship for proper attitude for flight and turn it over on its back to keep the horizon in sight. Humans can take over if the computers fail. The commander does nothing, however, unless that happens, leav-

Computers at Mission Control at Johnson Space Center, Houston, Texas, monitor the shuttle throughout each flight.

ing it all to the computers. But he or she is fully aware of the current conditions and is ready to take over in an instant if necessary.

As the shuttle climbs into space, everything happens automatically, under computer control, accommodating even high winds. Meanwhile, back on earth, computers are monitoring the flight and informing the astronauts of important moments as they happen, such as separation from the booster or achievement of orbit.

Although a shuttle pilot can safely land his or her craft back on earth, the computer is used for most of the return trip, and for minor corrections on the way down. The computer actually works with the pilot, to convert his or her decisions into precise and accurate changes in thrust, attitude, and direction, much as jetliners are now controlled. It's like having power steering for your car, only magnified a hundredfold and intelligent enough not to oversteer in a skid.

The computers that control the flight cannot be left to monitor themselves. They are more precise than human control but less adaptable to unknowns. On the trajectory plot a kind of cursor, a box-shape, moves continually, showing where the shuttle ought to be. The commander pilots the craft so that a symbol that looks like the shuttle is superimposed on the box, showing where the shuttle actually is. Again, the computer handles most of the work, with the commander having the final word, making sure things are working properly. Navigation signals from the ground help in the final approach. When the craft is subsonic, the commander takes manual control. The actual landing is handled manually.

When the *Columbia* returned from its third test flight, tracking planes with special infrared equipment on board sought to get images of its underside, to study the heating effects of the atmosphere. Onboard sensors could sample only a limited number of spots, but the tracker, in theory at least, could view the whole surface. As it was, only half the

underside was actually observed. Computers reduced the data to a false-color image, to see what parts of the shuttle became the hottest. This information turned out to be very valuable.

* * * * * * * * *

Whether we send a person, a robot, or just a TV camera into space, we have to keep track of where it is, send instructions to it, and receive the data back from it. This is long-distance communications in the extreme, and there are considerable problems to be surmounted to put it into practice.

Sometimes even the best-designed system overlooks something. *Skylab* was not supposed to fall back to earth the way it did, or when it did, but unforeseen solar activity created a bulge in the atmosphere, through which *Skylab* had to pass. This slowed the huge satellite enough to start it on its downward spiral. Scientists tried to correct *Skylab*'s decaying orbit without success; later they tried to control its descent so it would fall safely. They eventually succeeded in this, of course, but there was a special problem to be solved. *Skylab*'s onboard computer didn't want to respond to commands that went against its programming.

CHAPTER 6

THE FAR EXPLORERS

The first part of any science is observation. After that comes theory, experiment, and further observation. This is true for astronomy and other branches of space science as well, though sometimes it is not possible to experiment, since we can't bring a star down to the laboratory or make changes in planetary orbits. This puts a greater premium on theory, and especially on observation.

But observing the universe from the surface of our world has limits and sometimes is not even possible. That is why we explore—send satellites into orbit and spacecraft to the planets. In a certain sense, exploration substitutes for experimentation.

Satellites, spacecraft, and telescopes exist for the purpose of observing the universe and gathering information. This is, after all, what space science is all about. Only after much research of an observational nature has been done can practical missions be launched. And even these have, as their primary purpose, the gathering of more data.

If the mission is to another planet or the moon, control is needed along the way. The satellite or probe is affected by solar wind, micrometeorites, the sun's gravity, light pressure, and other conditions that are possible to predict statistically, but not precisely. As these conditions affect the motion of the vessel, course corrections must be made, and this requires that we be able to communicate with the craft. Communication is also needed to receive the data collected by experiments in space or on other planets.

ROBOT EXPLORERS

The space shuttle and earth satellites can be controlled to some extent remotely, but the *Voyager*, *Pioneer*, and *Viking* spacecraft, and other craft that go to the planets must have onboard guidance systems. For example, the time delay for a signal to go from the orbit of Jupiter to earth and back again is far too long for any kind of "direct remote control," to coin a phrase. It is not possible for someone on earth to look out through the spacecraft's viewports, see something happening, then turn a knob or throw a lever. By the time the instruction reached the spacecraft, the event would long since have passed.

Thus, a spacecraft is a kind of robot, in a very real sense. Because of the time delay for signals to reach it from earth, everything on board has to be more or less autonomous.

Some devices could be controlled by "clockwork," that is, with mechanical or electronic devices that do not require any computer functions. A soil sampler, for example, could be designed so that it began to function when a timer went off, and then operated mechanically, without any regard to the kind of soil, the angle of the surface

The Viking I *lander with soil sampler in action, on Mars in February 1977*

relative to the craft, the presence of rocks, and so on. But such a device would more often than not fail, and in a Mars probe, one usually has only one chance.

Instead, onboard computers analyze the situation and control the soil sampler's movements. The computer must constantly check the soil sampler's position to prevent it from trying to scoop up a sample while still above the ground or drive itself in too deep. If the scoop hits a rock, it must either move to one side or try to move the rock to one side. This is the same kind of decision making you perform unconsciously whenever you reach out for a glass of soda. Getting a machine to perform as well as your hand is not trivial. And it is the software, rather than the hardware, that makes the difference.

Soil sampling and all other functions of this type are handled autonomously, that is, the computer directs them without human intervention. To achieve this degree of autonomy requires considerable effort on the part of the designers. Almost every part, function, and action of a satellite or spacecraft must be able to withstand a range of conditions, must be able to work in more than one way, must be capable of remote correction, and must be backed up with auxiliary systems. Getting all this into the relatively tiny package that is even our largest spacecraft is not easy.

Unfortunately, any system that could do the whole job would be far too large. Instead, we have to content ourselves with small computer systems that can operate by themselves for a while but that periodically communicate with earth for reprogramming and corrections.

Such computers—and their programs—are usually able to accommodate changing conditions and minor accidents. Major disasters, such as being hit by an asteroid or the freezing up of a camera platform, usually require direction from ground-based computers under human control or sometimes direct human commands.

SPACECRAFT AS EXPERIMENT

Every spacecraft is an experiment, composed of many smaller integrated experiments. Much testing can be done on earth, in the laboratory, but the ultimate test is whether the device will actually work in space or on another planet, with the temperature, gravity, and other conditions that are in effect there, after its long trip through space.

The whole complex of experiments is computer guided and controlled, and this means that programs must be written. Not only must each one-of-a-kind computer configuration be totally reliable, so must the software that controls it. In one case, a programmer left a negation sign (a symbol like a minus sign but which means "not" instead of "minus") off one variable in a program, and the launch wound up in the Atlantic Ocean instead of in orbit.

One of the most important aspects of this kind of programming is the ability to respond properly to feedback. That is, the system must be aware of the conditions that affect its operation and be able to make adjustments as those conditions change. A telescope can be set up to move so that it follows a star by clockwork, but the greatest resolution is achieved when the scope is able to detect when the image is off center. Solar panels and heat radiators must constantly be realigned for maximum efficiency. Clockwork won't do; here, intelligent control by computers is vital.

It is the ability of intelligent systems to observe their surroundings that makes real success possible. A camera by itself only takes a picture, but a computer can make certain evaluations on the quality and content of that picture and make decisions based on those evaluations. It is, of course, their decision-making capability that distinguishes computers from other machines and that makes them seem intelligent.

In fact, it is such moment-by-moment control, response to feedback, and awareness of the surrounding environment that makes modern spacecraft by far the most sophisticated robots ever built. As marvels of mechanical and electrical engineering, they are probably unsurpassed. As examples of artificial intelligence, they are seldom appreciated. Instead of trying to simulate a human who is playing chess or making political decisions or acting as a psychiatrist, these robots simply try to imitate an insect, and they succeed remarkably well. If our cars and trucks were as well controlled, we wouldn't have to drive them. They could do all the work for us, and accidents would hardly ever happen.

Although people interested in artificial intelligence for its own sake are exploring certain possibilities, space scientists are concerned with practicalities, and if the computer can make a device or system function in a broad range of circumstances, it is, in effect, intelligent.

COMMUNICATIONS WITH HOME

As we have mentioned, the spacecraft on its way to Mars or past Jupiter is a complex robot. Of course, there isn't much for it to do but coast for a large part of its journey, but it has to keep telescopes trained on certain stars for navigational purposes and keep its antennas beamed at earth; these tasks it does itself, without direct human intervention.

Though a *Viking* or a *Mariner* is a robot, in one way it is quite different from most other robots. In the commonly accepted idea of a robot, everything is contained in one package, with at most some sort of remote controller "connected" by radio or sometimes physical wires. For a spacecraft, the communications components are scattered around the globe, the radio connection is extremely tenuous, and the whole system spans the distance between the earth and the spacecraft—tens or hundreds of millions of miles.

Reliability of transmission and reception are highly important. Even as a signal is being sent, it is checked for accuracy and corrected if necessary, and when it is being received, further checks are made. Often, incoming data can be corrected if it is known to be scrambled.

The signals sent must be coded in such a way that when they are received by the tiny craft, they can be accurately read. As the craft moves, its antennas must be adjusted so as not to lose the signal. The signal, when it arrives at either end, is usually very weakened. Communications theory is used to convert messages into signals that can't be misinterpreted and that can be translated by the craft back into information. The communications problems to be solved here are not trivial.

Once the observation has been made or the experiment performed, computers convert the raw data into binary signals, which are sent back to earth. For remote systems, conversion to binary signals is a vital step. Normal television broadcasting methods would not work well enough; television signals are too complex. Instead, everything is converted to binary data in such a way that even if a portion of the signal is missed or misinterpreted, the rest can be reconstructed.

Although sending instructions to a craft may require only short bursts of transmission, receiving data from a craft, on the surface of Mars or going past Jupiter or whatever, requires immense amounts of transmission time. With the aid of computers, transmission techniques have been developed that permit extremely high transmission rates, also aided by computers, so that the onboard devices can send back the immense amounts of data they collect. This data, in turn, requires computers to decode and interpret it.

THE DEEP SPACE NETWORK

In order to accomplish all this communication, ground stations need to know where the craft is. This requires incredibly accurate measurements, not only of the angle of the signal received but also of the time for a signal to go to the craft and come back again, and of the Doppler effect caused by the craft's receding from earth. Only computers can handle calculations and measurements this fine.

The most elaborate sensing devices in the world will do us no good if the information can't get back to us quickly and reliably. Even more important is trying to put a spacecraft in a precise place at a precise time, such as in orbit around Mars, so that a lander can actually touch down. And this can't happen if we don't know where the spacecraft is, within just a few hundred kilometers. The problems of communicating with a spacecraft so far away, which also involves knowing just where it is, are large and complex.

The system that handles all this is the Deep Space Network, or DSN, operated by the Jet Propulsion Laboratory and directed by NASA. The DSN is composed of radio telescopes at a number of places around the world and the rather complex communications channels between each "earth station." To accurately range (discover the position of) a craft, we must first know the correct measure of the astronomical unit (the average distance between the earth and the sun). The value astronomers had once assumed to be correct was found to be wrong when spacecraft first traveled beyond the earth's orbit, and had to be recalculated using data from those craft. The distances between ground bases must then be accurately determined to within a few meters. These two separate problems have been solved by sophisticated measurement techniques, so that right now, we can range a spacecraft with an error of 100 kilometers (62 miles) or less at the orbit of Jupiter. This is equivalent to shooting a bullet from New York to Los Angeles and hitting a target half a meter (1.5 feet) wide.

The whole DSN system is under computer control. And aside from its components, establishing the system required computer assistance in locating the best places for each station (political considerations were another problem) and in designing the communications links necessary between each one.

To provide this service, a vast number of problems had to be solved, in choice and design of materials, designing the equipment, developing communications theory, and so on. These problems were so large and complex that computers had to be used to help solve them, too.

Systems other than the DSN are used to communicate with spacecraft in earth orbit and between the earth and the moon, though some of these share facilities with each other and sometimes with the DSN. In many cases, the principles of communication are the same, though the DSN must meet more exacting specifications, due to the greater range of interplanetary spacecraft.

CHAPTER 7

VISIONS OF EARTH FROM SPACE

THE LANDSATS

Earth is the body that has been the subject of most human observation from space, and the satellites most famous for studying the earth from space are the *Landsats*, which orbit the globe every 103 minutes taking pictures.

The *Landsats* use four cameras to take pictures in four colors: yellow, red, short infrared, and long infrared. These pictures are recorded digitally and then transmitted to earth where another computer reconstitutes the data into pictures, using false colors.

Pictures from space tell us things about the nature of surface vegetation, minerals, human populations, and so on.

In the future, *Landsat* satellites will record in other wavelengths beyond the range of the human eye. These

A photo of New York City and environs taken by Landsat-4

images, reduced to digital signals, will be fed without human intervention directly to ground-based computers, where they will automatically be analyzed to produce drought warnings, forest inventories, land-use records, crop forecasts, flood alerts, earthquake predictions, and resource summaries. As is done today, the information these satellite images reveal will be combined with information gathered by people on the ground; for example, if a *Landsat* detects an area with a peculiar color, a ground survey team will be sent to find out what it means.

Landsat photographs can be analyzed for reflected color. Different surfaces, such as granite or sand, or even different types of vegetation, reflect light differently. In some cases, the actual color differences may be so slight that we can't see them at all with the naked eye. If we want to see them, false colors can be used.

SEASAT

Even when a satellite fails, as *Seasat* did only three months after it was launched in 1978, we can learn a lot from the data it does manage to send to us. *Seasat* was designed to examine the ocean floor by using a radar altimeter. This device beamed radio waves, which are unaffected by water, down into the ocean, where they were reflected by the ocean floor back to the satellite. The delay before their return was measured, as it is in any radar measurement test, thus indicating how deep the floor was at various points. (Actually, it was how far below the *satellite* the floor of the ocean was, but since the distance of *Seasat* from the center of the earth was known, it amounted to the same thing.)

The data at the time seemed interesting, but not all that useful. Then recently it was examined again. This time, the scientists processed the data using sophisticated computer-imaging techniques. The results were rather complete and incredibly detailed maps of the ocean floor.

Radar observations from space are used in other ways, too, employing a computer to analyze the data and, if desired, create pictures. Synthetic aperture interferometry of radar signals is used to give a picture of the shape of the earth and information about the kinds of minerals and geological structures on its surface.

Either a plane or a satellite can be used as the radar transmitter. A signal is bounced off the earth, forming an image, and a moment later, after the plane or satellite has moved a bit, another picture is taken. The effect is the same as if two radar devices were used instead of just one. Computers, of course, are necessary to later coordinate and combine the data to produce a meaningful image. The delay of the signals, as they are reflected from the surface of the earth, tells how far away the surface is. But the plane or satellite is moving, and thus the signals coming from a spot ahead of the craft are higher in frequency, and those coming from just behind are lower in frequency—i.e., Doppler-shifted. By combining the delay and Doppler-shift information, a true two-dimensional image can be produced.

NIMBUS AND MAGSAT

Nimbus 7, launched in 1979, maps sea ice, measures the concentrations of aerosols in the atmosphere, and measures carbon monoxide, methane, and ammonia. It can also measure the color of the ocean, as it shifts from blue to green, and hence determine the amount of phytoplankton in the water. *Magsat*, a magnetic-field satellite, took more than a billion readings in 1979 of the strength and direction of the earth's magnetic fields. These satellites don't necessarily produce pictures, but the data can be converted into visual images if desired.

FORECASTING THE WEATHER

Many evening news programs make use of radar pictures when they report on the weather. These pictures are produced from data received from the GOES weather satellites. Much of the information received is not converted into pictures, of course, but is used to help meteorologists tell what the pictures mean.

The use of computers is indispensable for serious weather forecasting. The National Weather Service uses fourteen computers to correlate data from four satellites, other data-collection platforms, and about seventy radar stations nationwide. These computers work in tandem, using the data received from all those sources to construct a mathematical description—a model—of our atmosphere. This model, in turn, is used to project the weather. The National Weather Service generates about 2,000 weather reports each day.

The weather forecasts based on this data depend in large part on statistical analyses. No model thus far developed can truly mirror what really happens in our atmosphere, so the theories of probability must be used. Long-range forecasting, predicting what will happen not tomorrow but next year, requires even more data and more calculations.

Students at the University of Colorado in Boulder used computers to sift data taken from the Solar Mesosphere Explorer (SME). This satellite orbits the earth at 336 miles (540 km) above the ground and is the only satellite run by a university and students rather than by NASA (though the Amateur Space Telescope will be another). The SME gauges seasonal and geographic variations in ozone and measures nitrogen oxides, water vapor, and the OH ion, all of which have an effect on the ozone layer, which is 15 to 50 miles (24 to 80 km) above the earth.

The students have written their own programs, as a part of their studies. The satellite also provides information on how the sun produces ozone, and it monitors for abnormally high bombardments of solar protons, both automatically, under its own computer control. These systems and studies are not new or experimental, but they are privately rather than publicly funded and are giving students first-hand experience at solving real space science problems.

MODELING THE EARTH

All the images of earth taken from space are processed by computers on the ground, long after the data is sent. Synthetic aperture radar interferometry makes use of time delay and the Doppler shift, as we have noted, to provide meaningful information about the surface of the earth, but first the raw data has to be processed. Even the Cray-1, one of our fastest computers, is estimated to be an order of magnitude slower in processing the data than is necessary for real-time imaging. That is, one minute of data can take up to ten minutes to produce a picture, depending on the resolution required.

But to take full advantage of the data, the ratio is more like 500 to 1, that is, one second of data would take over eight minutes to process. The *Landsat* satellites now take their pictures primarily in the visible or near-infrared portions of the spectrum. As new devices are developed, they will record other wavelengths beyond the range of the human eye. All these images, however they are recorded, will be analyzed by ground-based computers, without need for any human intervention.

Computer-controlled detectors can also gather information on gravity, and, although this doesn't produce a picture, it can be used to modify a picture or at least add to our knowledge of a physical system. For example, observations from space of how satellites subtly change their position from ideal orbits tell us about the shape of the earth, and about mass distribution within it.

In this way we can construct models describing the shape of the earth and the distribution of landmasses in its crust. We can combine the information taken from space with measurements taken on the ground to help us learn more about the movement of tectonic plates, predict earthquakes, or describe the shape of the continents in the past and in the future. Models cannot be constructed, of course, without some understanding of the underlying principles and laws that govern them, and not until sufficient data has been recorded and sometimes analyzed in other ways.

For example, by knowing how the earth and moon affect each other gravitationally, and how the moon's orbit is even now changing, we can run the model backward to try to discover if the earth/moon pair formed together, or if the moon was once an independent body captured by the earth. We can also predict whether the moon will come closer or farther away from the earth in the future, how long it might take before the moon collides with the earth or escapes from its gravitational field, and what the effects on the earth might be, in either case.

Sometimes, special techniques are used. For example, though the moon shines with reflected light, sometimes it's better if we shine a little light of our own on it. By bouncing laser light off the moon, we can gather information about how it is moving, vibrating, and changing its position. A rather special photoflash, indeed.

COMMUNICATIONS SATELLITES

Perhaps one of the most important developments of space science has been the communications satellites that link the whole world in a global network. In large part, once these satellites are in orbit and their antennas are properly aimed (by computer as well as by remote control), there is little further need for them to be controlled, except when minor perturbations in their orbit or orientation require action.

But all communications satellites must be autonomous. All signal receiving, switching, and transmission must be controlled by onboard devices. With only a few channels to accommodate, electromechanical devices could do the job. But as the world comes to depend more and more on satellite communications and the number of channels increases, the complexity of the signals sent also increases, with computer control of these "telephone" functions becoming more and more necessary. Most telephone signals are analog, but this method is inefficient and subject to certain kinds of transmission errors. In the future, all satellite communications will be digital.

Sophisticated communications are very important to today's society. Without our global satellite network, even common everyday business might be disrupted.

In 1974 and 1975, geosynchronous communications satellites helped prevent a worldwide banking crash. Inflation, economic differences between developed and undeveloped nations, and a fourfold oil price increase contributed to the problem. Banks were in danger of collapsing and thereby setting off a depression because of a cash-flow problem in the weaker companies and countries. (Indeed, this condition persists even today.)

Just to keep money flowing, it was decided to "recycle" OPEC's oil profits, and to do that, international bankers used computer-to-computer satellite facilities, known as trans-border data flows, or TDFs, which are capable of moving 25,000 pages of data around the world in minutes.

How the world money market actually juggled the cash is one thing, but without the computer, it would have been

An artist's concept of an RCA communications satellite

impossible. Communications satellites link banks to each other regardless of borders. They are more or less independent of national or international regulation, which is a cause of concern for some governments, which would rather they retained control.

Similar computer-controlled global communications will eventually have an effect on the decentralization of manufacturing. Engineers in one country will be able to share their work with those in another without physically being there. This will save on travel expense, time, accommodations, and, perhaps most importantly, the necessity of having to coordinate everything to happen at the same time. With teleconferencing, messages can be relayed to other people to be read at their convenience. In practice, this amounts to combining the best qualities of physical mail and telephones. The Ford Motor Company estimated that designing the Escort using computer communications saved the company $150,000. Ford's distribution and sales are also facilitated by satellite communications.

Intelsat provides communications between places in countries which, although not that far apart by American standards, have no physical link, no phone lines. We can call New York or Los Angeles over phone lines, but in many Third World nations such lines don't exist, and are either too difficult or too expensive to install. Here, the satellite is the only means for some people to communicate with their neighbors.

Computer-controlled communications, facilitated by computer-controlled satellites, are thus one of the most important products of space science. And, when decisions have to be made by humans who are widely separated, teleconferencing is proving to be an answer. To provide this service and coordinate it, computers are the best, if not the only, tool.

CHAPTER 8

THE SOLAR SYSTEM

Though the word *astronomy* comes from a word meaning "star," astronomers don't limit their investigations to faraway stars. The study of our sun and the solar system's planets and moons is also a part of the science. Of course, all of the imaging techniques mentioned so far are also used to examine these bodies.

Observation comes first, in any scientific endeavor. Because we can't measure planets or their orbits directly or physically, we must use statistical methods to analyze many hundreds, thousands, or millions of indirect or partial measurements. Here again, only computers can handle the vast quantities of data received, whether the results produce a picture, a graph or table, or just a number.

MONITORING THE SUN

The Solar Maximum Satellite examines and monitors the activity of the sun. A Jet Propulsion Laboratory instrument

Solar Max, shown here in the Shuttle cargo bay, was repaired in space and is back in service gathering data on the sun.

on board once produced data that a computer used to find a link between sunspots and the sun's luminosity (brightness). As sunspots increase, luminosity decreases—in this case, by 0.2 percent. This may not seem large, but a typical fluctuation is only a few hundredths to a tenth of a percent, and it doesn't take much of a reduction in luminosity to have a noticeable effect on our weather.

James A. Barnes has developed a computer model of sunspot activity for the National Bureau of Standards. This model tries to show that trends in the production of sunspots arise in an unpredictable fashion. His theory is that the sunspot cycle is caused by processes within the sun, and that these same processes send random signals to the surface. The sun, according to this theory, ignores some of these signals but reacts chemically to others. This has a filtering effect which, in turn, causes the sunspots.

Using this model, Barnes produced a computerized simulation of the complex process, basing it on the eleven-year cycle of sunspot activity. The computer produced a graph that was basically noise but that had a signal with regularly occurring highs and lows that simulated the observed sunspot pattern surprisingly well.

Portions of this graph were compared with one that depicted 250 years' worth of sunspot activity. Both were found to have an almost identical pattern, even though the individual events differed. The simulation even had Maunder Minima (periods when there are no sunspots at all) and Grand Maxima (times when sunspots are unusually large and active).

In a similar way, other astronomers using data from the earth-orbiting Interplanetary Monitoring Platform Number 7 (IMP 7), which monitors and samples waves in the solar wind, prepared models of these waves. *Pioneer 10*, on its way to Jupiter, also provided samples. These samples, as data, were then compared with actual observations. In this way, astronomers hope to discover the overall pattern of the solar wind in the plane of the earth's orbit.

A LOOK AT VENUS AND MARS

Radar has been used to "look" at the surface of Venus, which is shrouded by clouds. Except for a few pictures from Soviet venerean landers, this is the only way we have at present to learn about the geological structures of the planet. The resulting images are printed in false colors, the differing shades of which indicate the altitude of the surface above the planet's center.

The TV cameras that brought us pictures of the surface of Mars were a kind of robot. Rather more sophisticated was the apparatus within the lander, which took the samples from the soil scoop and tested them in several different experiments to try to detect traces of life. Then, as the data came in, it had to be converted into a form that could be transmitted to receivers on earth. The presence of a particular gas that showed up during one of the experiments, for example, was noted by detectors, which counted the molecules and converted that count into binary digits. The data from the other experiments—measurements of light intensity, wind velocity, atmospheric pressure and composition, temperatures, and so on—all had to be converted into digital data. This was done by special analog-to-digital translators, and computers also stored and transmitted the data.

PIONEERS AND VOYAGERS

Models are mathematical descriptions of systems or processes. Information about these systems and processes, plus our knowledge of scientific principles and laws, provides us with the data necessary to construct these models. Gareth P. Williams of Princeton University constructed a mathematical model to try to account for Jupiter's east-west bands of colored clouds, based on ground observations and on data transmitted from *Pioneers 10* and *11*.

Computer imaging of the rings of Saturn

The characteristics Williams used to describe these belts and zones, and the Great Red Spot, included infrared measurements, cloud height, vorticity (degree of spinning), pressure, temperature, vertical velocity, and color. The result, of course, was a rather elaborate and complex model of the weather high in Jupiter's atmosphere. In a similar fashion, by comparing theories of atmospheric movement with observations of the Great Red Spot, computer models were constructed to try to explain its dynamics.

The high-resolution cameras on *Voyagers 1* and *2*, which took pictures of Jupiter and Saturn, were controlled by carefully programmed computers, which were not only able to direct the cameras according to the scientists' instructions but could also accommodate, in real time, the changing conditions in the space environment near the planets, and the changing positions of the moons and planets relative to the spacecrafts as they went by. Special feedback control held the cameras steady in order to get good pictures, fifteen times steadier than the movement of a clock's hour hand.

When *Voyager 1* swung by Saturn in November 1980, it detected some strange electrical signals, which proved to come from the rings rather than from electrical storms on the surface, as had first been surmised. These signals were subjected to sophisticated computer analysis, which revealed a cycle of approximately ten hours and ten minutes—confirming that the signals could not be coming from the surface since that rotates at a different rate. A large moonlet within the rings was proposed as the cause, but this has not yet been substantiated.

Voyager 2 did, however, discover a mysterious 100-yard (91-m) gap in the rings, right where the signals should have been coming from. Such a gap could have been made by a moon sweeping up material as it orbited there. However, by the time the spacecraft made its observations, the nature of the signals had changed. There were fewer electrical discharges than before, but the power of each discharge was the same. Various suggestions as to the cause of the signals are all highly speculative.

URANUS'S RINGS

Perhaps one of the most exciting results of computer analysis of optical information without the production of a picture was the discovery that Uranus has rings. In 1977, NASA's Kuiper Airborne Observatory (KAO), a modified C-141 jet airplane designed to fly above 75 percent of the earth's atmosphere, observed an occultation of the star SAO 158687 by Uranus. To make this observation it carried a telescope, a photometer, and lots of computer equipment. KAO was supposed to measure the time it took for the star's light to dim as it passed through Uranus's atmosphere, which it did, but at the same time it accidentally discovered the rings. The existence of the rings was determined by a computer analysis of dips in the star's light before it touched the limb of Uranus.

LOOKING FOR ASTEROIDS

Photography is used extensively to look for new asteroids. One type of asteroid, the Apollo type, which travels relatively close to earth's orbit, is of special interest. Most Apollo asteroids are very small indeed, and the photographs have to be taken with long exposure times, which results in a picture with thousands of points of light. Almost all of these points are background stars. The near stars are easily distinguished because of their brightness, but the farther stars do not look much different than the image of an asteroid.

However, an asteroid can be detected because the telescope is made to follow the stars, so that their images are points, while the asteroids, which are moving at a different rate from that of the stars, produce images that are streaks. But these streaks are usually very short and very faint, and finding one among thousands of points is a literal hunt for a needle in a haystack. Examining the photos for the telltale streak takes a long time, and an asteroid can be easily missed.

To help in the discovery and study of asteroids, a new scanning system is being designed at the University of Arizona, with the help of NASA. Work on this got under way in 1981 and may soon be finished. The basis of this system is simple digitization; that is, pictures will be taken of the sky, and the images will then be translated into arrays of numbers, each of which will describe the brightness and position of an object in the sky.

The computer does not work with the photos themselves but compares the number arrays. The program notes any unaccounted-for changes between two pictures taken just a few minutes apart. If the data were simply converted into pictures, their short exposure times would be unlikely to reveal any except the largest, nearest, or fastest asteroids. Humans, examining the pictures, would see only points of light and could not distinguish an asteroid from a distant star.

But the computer can note even tiny differences, and pinpoint them exactly. Once such a difference is detected, it can be brought to the attention of an astronomer, who can then examine that portion of the sky more closely by using longer exposure times or more powerful telescopes.

TRACKING COMETS IN THE SKY

That computers can be better than humans at tracking objects in the sky can be demonstrated by a case in which the mechanical wheels that turned a telescope slipped, and control had to be maintained by hand—literally.

At the McDonald Observatory of the University of Texas, an IBM 1800 was being used to track Halley's Comet. At that time, the comet was not visible, so the telescope was aimed at a star near where the comet was supposed to be. Photographic plates had to be exposed for many minutes in order to pick out as faint an object as the comet was then, and the direction had to be controlled with a guiding error of less than .15 arc seconds in order for the image of the comet to be distinguishable from any asteroids that might happen to be in the field of view. This control had to compensate both for the earth's movement and for the comet's computed movement; otherwise, the image would have been a streak instead of a point.

But the turning wheels that were controlled by the computer were observed to be slipping by as much as a full arc second per hour. The astronomers had to revert to a manual system, which involved looking through an eyepiece, aimed at the star near where the comet was supposed to be, and moving the telescope by turning control wheels by hand.

But humans make mistakes. Instead of turning a thumbscrew a number of arc seconds per turn, the astronomers made a number of turns per arc second. The value for rate of control was simply inverted. As a consequence, the telescope moved 50 percent too fast, and no useful pictures were obtained at all. Human error, fatigue, and the invisibility of the object sought all contributed to this failure. Had the turning wheels been working properly, so that the computer had control of the telescope, they might have learned something. As it was, their opportunity was wasted.

The astronomer Fred L. Whipple and his associates developed a computer model to explain the movements of the Comet Enke. They based this model on the idea that a comet is what Whipple calls "a dirty snowball," that is, a mass of ice and frozen gas containing large amounts of dust and other nonvolatile material. Such a body, when heated as it nears the sun, would emit jets of gas that would make it spin, change its orbit slightly, or both. These orbital changes would be independent of perturbations by planets or the pressure of the sun's light (the solar wind).

In running this model, the astronomers used two computers, each programmed separately and with a different coordinate system, as a kind of double check. The computers calculated the rate of precession of the spin axis of the comet (that is, the degree to which the comet wobbled like a top) and the nongravitational orbital forces (such as jetting gas and solar wind) acting upon it.

The results of their simulation corresponded with their observations of the actual comet, though they had to make a few corrections to their model. Nevertheless, the model validated the theory, predicting the direction of the fans of jetting gas and supporting the idea that the comet was made up of snow (though not necessarily water-snow). It helped calculate the size of the comet's nucleus, its density, its probable shape, its rate of spin, and the rate at which mass would be lost as the gases boiled away.

A TENTH PLANET

Although the gross orbits of planets can be calculated by hand, the mathematics is rather complex, and it is much easier to use a computer, not to mention an awful lot faster. This ease and speed becomes more important when astronomers want to calculate the orbits of more distant planets—those beyond Saturn—and when they want to take into account the perturbations (disturbances in orbit) caused by the gravitational pull of one planet on another. The classic example of orbital analysis was the discovery of anomalies in the orbits of Uranus and Neptune, which eventually led to the discovery of Pluto. Further study, however, combined with recent data on the mass of Pluto, and Charon, its moon, convince most astronomers that Pluto could not in fact produce the perturbations in either

Uranus's or Neptune's orbits. There is strong reason to believe, because of this, that Pluto is not the only planet out there.

Thomas Van Flandern, of the U.S. Naval Observatory, is one of those convinced that there is a tenth planet beyond Pluto. He used an IBM 4341 to match certain assumptions against a huge quantity of observational data about the orbits of Uranus and Neptune. These assumptions were educated guesses as to how big a tenth planet might be (two to five earth masses), how far from the sun it was (50 to 100 astronomical units), and what its orbital tilt was (very tilted and quite elliptical, even more than Pluto's orbit). From the results of this matching, he and his team obtained a list of places to start looking for the hypothetical tenth planet. Such a model cannot predict the one place where such a tenth planet might be, but it can eliminate most of the places where it most certainly will not be—given, of course, that our data on planetary perturbations is correct.

As *Pioneers 1* and *2* leave the solar system, their speed will be carefully monitored. Computers can detect changes as small as a fraction of an inch a second. If there is a distant planet of sufficient size to alter the motion of the *Pioneers*, it will almost certainly be nearer one *Pioneer* than the other, and its effect on the nearer *Pioneer*'s motion will be stronger. The other *Pioneer* might not be affected at all. By calculating the changes, if they occur, a possible location for the tenth planet can be determined.

CHAPTER 9

NEARBY STARS AND FAR PLANETS

Successful observation and analysis depend in large part on knowing what a particular form or strength of radiation means. If an object emits bursts of X-rays, does that mean it is pulsing inside, rotating, or that material is falling into it? This can be learned by performing experiments in the laboratory and then extrapolating to extraterrestrial systems, such as the sun, Jupiter, or distant stars.

For example, some astronomical and planetary objects produce energy in invisible wavelengths, but these energies are associated with well-known physical processes. Thus, if X-ray output is high, we know something interesting is happening and even have some idea what it is.

By observing our sun and comparing what we see with theories about how it should work, we can learn more about how it actually does work. We can get a good idea how old it is, how stable it is, and how its fluctuations affect us. Some of these effects on our planet are rather obvious, such as ice ages. Others are more subtle, as in the amount of radiation we get from the sun daily, the weather of our whole planet as well as of specific areas on its surface, the effect of the sun on radio communications, and so on.

We can create computer models of how a star works, to determine when the sun formed, how long it will last, and what changes it will go through as it grows older. Our sun is not constant, and by using models we can learn how its fluctuations might affect our lives. Some stars have special qualities, such as very short life spans or extreme size, and we can learn where to look for stars like these. Such studies also tell us more about the kinds of processes that go on in our sun and other stars.

Sometimes, we can learn more about an astronomical object by seeing it in more than one area of the spectrum at a time. For example, astronomers frequently compare a radio image with an optical one, or an X-ray image with an optical picture. Of course, X-ray and radio images can be compared with each other, too. This was done to study the star SS 433.

In conventional photographs, SS 433 is quite unimpressive. But when radio and X-ray data were taken of the star and fed to the computer, generating two images, one superimposed on the other, the star appeared to have two jets of matter being ejected in opposite directions at very high velocities. It is now thought that the star is really a supernova remnant, but its true nature, the reason for and mechanism of the jets, and other anomalies are still a mystery.

The process of superimposing one kind of observation on another can also be used, in some cases, to look at least a little way into the interiors of stars. We now know, from our observations of the sun and from the laws and theories of physics, that radiation is produced by certain processes, and that those processes occur at certain lev-

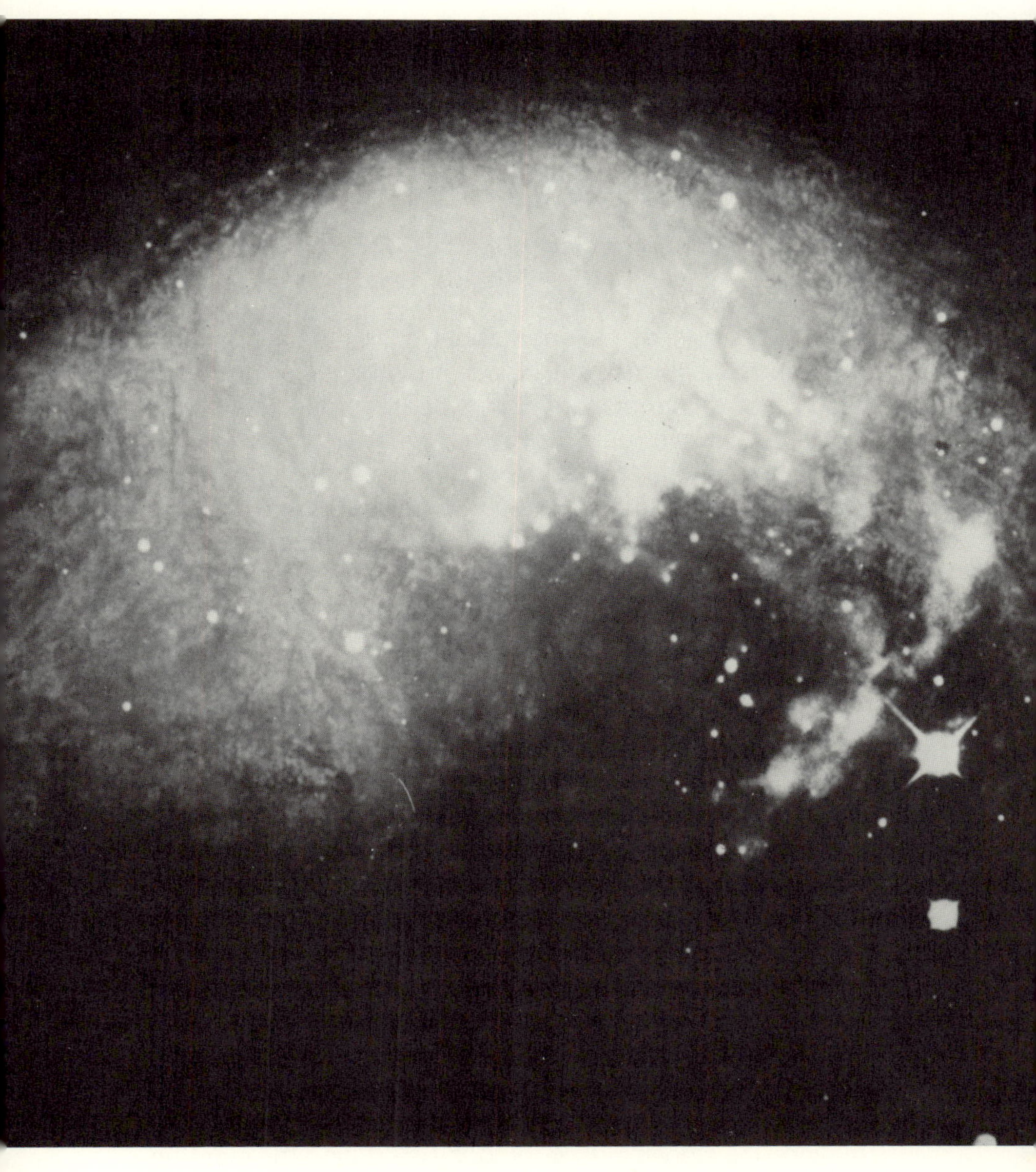

The Orion Nebula: an infrared image superimposed on a traditional photograph

els within a star. An ultraviolet picture of the sun is very different from an X-ray picture, or an infrared picture. We also know that the radiation that reaches us is modified by the structure of the star, and the way the radiation is modified tells us a lot about that structure.

OTHER PLANETARY SYSTEMS?

From what we have learned about our own solar system, we can make some guesses as to whether or not other stars might have planets. By comparing other stars with our sun, we can learn which are most like our own—in mass, temperature, possible age, rate of rotation, and so on—and hence which ones might be likely candidates to have planetary systems of their own. We can even make some guesses as to where we might find other life and intelligence in the universe. But detecting the planets themselves, until very recently, has been an almost insurmountable problem.

Part of the problem is that although we can see planets in our system by the light they reflect, those that might orbit another star would be so dim in comparison to the star itself that any light from them would be completely washed out. With the Space Telescope, it might be possible to put a shield between the star and the scope, as is done to view the solar corona, and thus mask off much of the starlight, enabling us to see the reflected light from the planet.

Another method of discovering planets involves measuring minute perturbations in a star's proper motion as it moves across the sky. (The proper motion is the angular change of position of a star in the sky, not its actual motion.) If a significantly large body is orbiting the star, the proper motion of the star, over a period of time, should be some form of sine wave, or regular oscillation from side to side, instead of a straight line. The problem here is that the inaccuracy of measurements is usually greater than the perturbations even a body fifty times the mass of Jupiter would produce.

Recently, the Infrared Astronomical Satellite, IRAS, has detected what astronomers think is a disk of dust surrounding Vega. This could mean that Vega has an infant planetary system. Further observations from IRAS and the Space Telescope will tell us more.

THE PLANETESIMAL THEORY

By making observations of the planets in our solar system, either from observatories on earth or from spacecraft, we can learn about their structure, the composition of their surfaces, and any atmosphere they might have even before we send spacecraft to explore them. From that, and from other observations and theories, we can get some idea of when the solar system was formed and how it was formed, whether by the accretion of millions of planetesimals (tiny protoplanets) or the sudden collapse of a dust cloud.

Scientists and astronomers convert their theories of planetary formation into mathematical formulas, then solve the formulas to see what the results are. If the model shows that planets will be distributed much as they are in our solar system, then the model can be assumed to be more or less accurate.

In one case, a computer simulation was created that started with a hundred planetesimals. These were randomly distributed, within a plane, at distances of ½ to 1½ astronomical units from a hypothetical sun very much like our own. Their orbits were given eccentricities of less than 0.15—that is, very nearly circular. Running this model showed how such bodies could eventually accrete into a number of full-fledged planets.

The simulation not only performed the calculations but also produced graphs showing how the original bodies moved closer to or farther from the sun, and how the eccentricity of their orbits changed as the bodies interacted with each other. But the model assumed that all the bodies would be in a two-dimensional plane, and most such bodies would actually have their orbits tilted with respect to that plane, moving in a three-dimensional universe. So, they tried another simulation, to take this orbital inclination into account.

This model produced similar results, though the process took (on various runs) from 1,000 to 7,000 times longer. In the two-dimensional model, the time estimated for the hundred planetesimals to aggregate into four planets was about 61,000 years. When orbital inclination was allowed for, it was estimated to take over 440 million years.

When planetesimals orbit the sun, they interact with each other in complicated ways. Their velocities with relation to each other determine whether they will hit or merely perturb each other. If they hit, they will either coalesce into one larger body or break apart, perhaps into a large number of smaller bodies. Large bodies attract smaller ones by gravity, in spite of their theoretically independent orbits. Various computer models of this theory have been tried, with varying results, but they have all tended to confirm the planetesimal origin of planets.

MODELING TO DISCOVER THE ORIGIN OF LIFE

Nearly any model, no matter how theoretical, can tell us something about how the solar system was formed or how it functions today. From what we know about this, we can extrapolate to the possible formation of other planetary systems around other stars.

In a similar fashion, we know quite a bit about our earth today, though by no means everything. Still, taking what we know, we can extrapolate backwards, to try to discover what conditions on earth were like hundreds of millions of years ago, or billions of years ago.

We can try to decide, from other studies, what conditions were necessary for the formation of life, and how likely such conditions are to occur on other worlds around other stars. At present, there are several contradictory theories about how life got started on our own planet, and models, as well as laboratory experiments, will help us decide which one is correct. Needless to say, such modeling requires an immense amount of calculation.

* * * * * * * *

In studying our solar system, we started with observations, then sent devices to the planets themselves, to perform experiments. In trying to learn more about other possible planetary systems, it would help if we could do the same—send spacecraft to stars of interest, to make more direct observations and take actual measurements.

Though traveling to other stars is not now feasible, several designs for interstellar probes have been proposed. Any such craft would require an "intelligent" onboard computer to control the robots that would maintain and operate the ship, and to direct the probe's signals back to earth.

One possible starship drive was suggested by Roderick Hyde and Lowell Wood at the University of California's Lawrence Livermore Laboratory. The starship would be powered by laser-induced microfusion.

A colleague of Hyde's and Wood's, George Zimmerman, wrote a computer program called LASNEX, which modeled and predicted what would happen in laser-fusion experiments. The model was found to be accurate. The results of actual experiments were very close to those predicted by the model.

By using such models, scientists will be able to eliminate unnecessary real experiments and concentrate only on those that produce desirable results. Those results, of course, can then be added to the data on which the program runs.

CHAPTER 10

THE MILKY WAY

Our star is one of a local group of stars. We can study the proper motions of all the stars in the group, and from that create a model to answer some interesting questions. For example, will another star soon pass near ours, or has one done so any time in the (relatively) recent past? Simple calculations involving just a few stars can be done by hand, but the more stars involved, the greater the need for a computer.

Because sophisticated imaging techniques can let us "see" almost any form of energy, we can learn a lot more about the nature of the universe and the structure of our galaxy than is possible with visible light alone. Interstellar gas and dust emit energy at certain frequencies and wavelengths depending on their temperature, density, and molecular makeup. By mapping the distribution of these clouds and identifying their component elements, we can determine the structure of the galaxy in whole or in part.

Sometimes, when thinking of all the things computers do, we lose track of the fact that they do compute. We see the pictures they produce, watch the space shuttle land, and see demonstrations of physical experiments performed on Mars. But in spite of all these advanced functions, computers are still often used just to perform mathematical operations on numbers, quickly and accurately, hundreds or thousands or millions of times a second. For example, they are used in solving mathematical problems before the spacecraft is built, as it is flying, and after is has accomplished its mission.

Whatever is eventually done with the raw data—that is, whatever form we finally see the results in—a large part of the processing involves mathematical manipulations. Producing pictures from data, or graphs and charts, is not a direct transformation. The computer must "crunch numbers"—perform mathematical functions—sometimes applying sophisticated mathematical theories to the data involved. What we see is only the end product of a long series of calculations.

Consider, for example, a sky survey taken to learn more about the structure of the Milky Way by mapping galactic dust and gas. More than 140,000 different positions in the sky were examined, recording the velocity of the dust and gas at one hundred different velocity levels. This produced 14 million data points, which were then manipulated to produce pictorial maps and synthetic photographs—that is, pictures produced solely by a computer, without a photographic plate ever being exposed to starlight. In the past, astronomers took a printout of the numbers and colored over each number by hand, each one getting a different hue and thus producing a crude kind of map. With this amount of information, however, such a task would be impossible.

Obviously, data in this form is not easy to understand,

even as pictures. Thus, various kinds of statistical operations are often performed on it, to provide graphs and charts, which can then be analyzed for meaning. The meaning of those graphs and charts can then be made even clearer by performing additional statistical operations.

Without a modern digital computer, all this would be next to impossible, and certainly not practical. Because of the vast amounts of data produced by our observational equipment and experimental devices, and even better ones being developed, computing is one of the major activities at most astronomical observatories today.

To make a complete visual mapping of the sky at the resolution desired would be impractical. Instead, the computer is used to construct a mathematical or statistical model. Some of these models are of the universe as a whole. Others are more limited in scope, such as those that map only the carbon monoxide in the galaxy.

A number of observations are made, with scientists then analyzing the data in a statistical manner, following what are called Monte Carlo techniques. The computer constructs a random model of clouds in the galaxy, following certain parameters the scientists give it, such as the size of carbon monoxide clouds, the speed of rotation of the galaxy, and the thickness or density of the clouds.

In order to make the models work, the astronomers also make some guesses concerning what they want to determine. For example, they might estimate the relative velocities of the clouds with respect to each other, the number of clouds in the galaxy, and how far apart they are. The computer then places these simulated clouds randomly in the model galaxy, which it then "observes."

Real observations are then made, and by comparing these with the computer's output, the astronomers can test some of their hypotheses. The model is adjusted by changing the assigned parameters until what the computer produces by statistical methods matches what is observed. From that, further assumptions about the model and about the rest of the galaxy are made.

FILLING IN THE BLANKS

Looking in certain directions, the clouds of gas and dust we've just talked about obscure what is behind them, leaving blanks in our maps. But certain forms of radiation can pass through these clouds, and by detecting these, we can fill in the gaps somewhat. We can also make inferences when we know that certain kinds of energetic emission are associated with certain physical or chemical processes, or with certain kinds of objects. If we know, for example, that a particular form of X-ray emission is associated with hydrogen at a certain temperature, then we can infer the presence of a hydrogen cloud, though we cannot see it directly.

Until recently, the core of our galaxy was invisible to us, not only because of intervening clouds of dust and gas but also because the huge amount of certain kinds of radiation coming from it blinded us to the other kinds of radiation it was emitting. But infrared astronomy has provided a solution here, so that now astronomers can examine the very center of the core. These infrared radiations are carefully measured, with a computer then generating a color-coded picture.

Cosmic rays can also be detected. Solid-state devices on the satellite ISEE-3, the Third International Sun-Earth Explorer, do just this, sampling particles that help reveal the structure of the Milky Way. The satellite also carries ten other experiments for observing and measuring a variety of other phenomena.

ISEE-3 is in an interesting orbit. It is 1½ million kilometers (.93 million miles) from earth, about one hundredth the distance from the earth to the sun. In this position, it is subject to nearly equal gravitational pulls from both the earth and the sun, a "Lagrangian point." Using small thrusters to continually vary its position slightly, it always remains approximately between the earth and the sun, sometimes slightly above the plane of the earth's orbit, sometimes below. If you could see it from earth, it would appear to be circling the sun.

* * * * * * * * *

Given a mathematical description of the number of stars of a given brightness in a particular region of the sky, a computer is able to show us what our galaxy might look like edge-on, as if seen from another galaxy. The image resembles those actually seen of other galaxies.

Similar statistical analyses can be made of the counts of stars taken over the whole sky. We do not have to count the stars ourselves. We can let the computer do it. And by counting galaxies instead of stars, we can begin to speculate on the structure of the universe as a whole.

CHAPTER 11

THE COSMOS

In our study of the galaxies, we can do no real experiments at all. We can only make observations, try to explain them mathematically, and then design models to see how well we can predict the outcome of certain events. Our studies start with observations of nearby stars (other than the sun) and interstellar space.

We can detect molecules in space, their kind and distribution. We can map the structure of our galaxy, or at least parts of it. We can observe unusual objects, such as pulsars, neutron stars, quasars, and (if they really exist) black holes. We can observe galaxies moving in relation to each other and learn which are associated and how they evolved as a group. We can see how we are moving relative to the universe as a whole and how the universe is moving relative to us. We can detect the overall energy content of the universe and guess at the mass, size, overall structure, and distribution of mass within the universe.

From all these observations we can create computerized models to tell us more.

COSMIC RAYS

When a cosmic ray enters the earth's atmosphere, it interacts with the molecules of air and emits photons that behave in certain identifiable ways. These tiny bursts of light, brief and incredibly faint, are called air-showers. No human eye, no ordinary telescope, can detect them.

But there is a device called the Fly's Eye that is an array of photomultipliers. It does not produce an image but can detect single photons. When a photon is detected, its behavior is also observed, and when the computer connected to the Fly's Eye recognizes this behavior as being that of an air-shower event, it triggers a high-speed motion picture camera, which then photographs the secondary particles resulting from the interaction of the cosmic ray with the molecules of air.

From these pictures, then, the astronomer can tell the direction from which the cosmic particle came, and the energy with which it entered our atmosphere. Thus, we can learn something more about the nature of the cosmos as a whole.

SUPERNOVA STUDIES

Astronomers search for supernovas in other galaxies by taking series of pictures and comparing them for differences. A real-time search program conducted at the Corralitos Observatory of New Mexico State University near Las Cruces, New Mexico, in 1976 and later, uses a 24-inch (61-cm) reflecting telescope in its search. It is programmed to move rapidly and automatically, observing hundreds of galaxies in a single night.

It uses a TV camera to photograph each galaxy for just a few seconds. This picture is produced on a TV screen, where it is compared with another picture taken of the same galaxy some time previously. The observer then decides whether or not a "new" star has appeared, and, if so, further study is made. An observer can check hundreds of galaxies each night and make decisions immediately after the picture is taken, instead of having to wait for film to be processed. If a possible supernova is detected, larger telescopes are used for more detailed examination of that part of the galaxy where it was observed.

UV spectrometers, used for studying supernovas observed with the 200-inch (508-cm) Hale Telescope on Mount Palomar, make use of an instrument designed by J. Beverley Oke of Caltech, a multichannel spectrophotometer that combines the precision of photometry with the detail of spectroscopy. There are thirty-two photomultiplier tubes inside the instrument that can rapidly measure the brightness of a supernova in each of a hundred or more narrow bands of wavelengths, ranging in width from 20 to 360 angstroms. The device can measure wavelengths from 3,100 angstroms in the UV to 11,000 angstroms in the near infrared. Data from this instrument, obtained by Halton C. Arp, Oke, and their colleagues, has been instrumental in providing astronomers with much of what they know about supernovas.

QUASARS AND OTHER RADIO SOURCES IN THE UNIVERSE

Though radio waves are invisible, computer-imaging techniques used by the VLA can produce "pictures" of these radio sources with a resolution as great as that from the best optical telescopes. Quasars are among such objects studied by the VLA, as are radio galaxies, which are elliptical galaxies that, optically speaking, look like nothing special. But in radio frequencies, these galaxies are

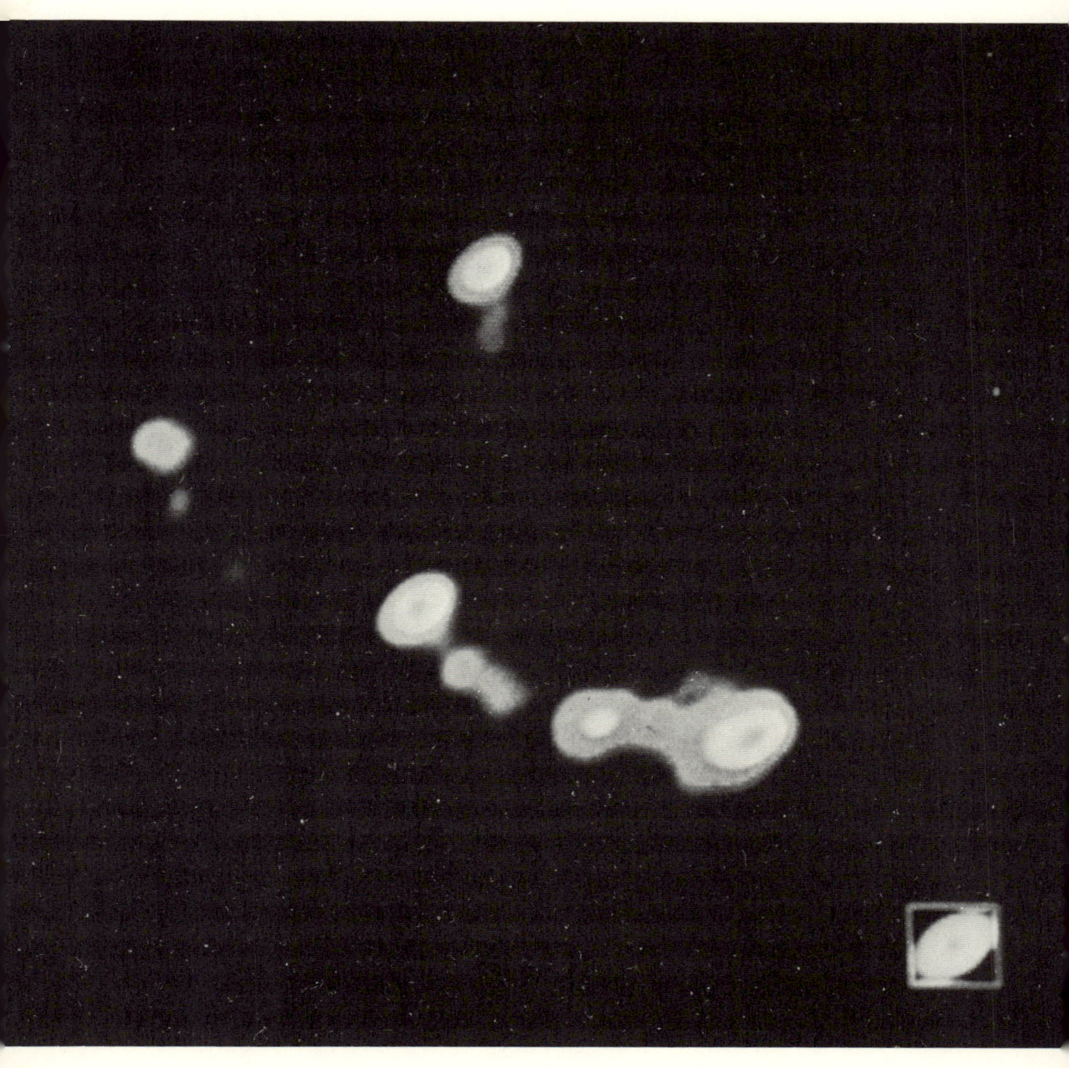

This radio map photo shows a quasar that has been multiply imaged by the gravitational lens effect. The two images of the quasar are the pinpoints of light north and south of center; the other images in the photo are radio emissions from the quasar that have not been multiply imaged.

extremely powerful emitters. The actual area of radio emission is many times greater than that for a normal galaxy, up to 170 times the diameter of the Milky Way.

A dramatic example of image enhancement was the discovery of the gravitational lens. Astronomers, using radio telescopes, were studying two quasars that were very close together. Pictures taken in visible light showed two optical images that, when analyzed, showed the same degree of red-shift. That two quasars should be receding from us at the same rate was statistically highly unlikely. That the two had identical spectra in other respects made it statistically impossible.

One explanation was that a galaxy lay between us and a single quasar, with sufficient mass to bend the light coming from the quasar and thus produce a double image. (In fact, there were three images.) While optical evidence alone was not sufficient to prove the gravitational lens theory, the fact that the radio source was also doubled tends to strongly confirm it. Both the VLA and the MMT were used in this study.

The original pictures did not show the intervening galaxy. Quasars are very bright objects, even though their great distance makes them look dim to us. This quasar was brighter than the galaxy between us and it, causing the galaxy to be invisible to us. But, by using a carefully designed computer program, the photographs were clarified just enough to show that indeed there was a galaxy where it should have been.

BLACK HOLES AND OTHER UNUSUAL OBJECTS

The use of computers to calculate and process data is not confined to professionals. Students at Caltech investigate unusual and hypothetical cosmic phenomena, such as black holes, white holes, quasars, neutron stars, wormholes, and tunnels through hyperspace. Some of these bizarre objects, such as black holes, we are pretty sure exist, though they have not been observed directly. Others, such as wormholes though space, are less likely to really exist, though they are based on theories that otherwise hold true.

Kip Thorne, professor of theoretical physics at Caltech, supervises these students, who are studying Einstein's equations and the theories of relativity. They need no telescope time, only sufficient computer time to work out the equations. The students study whether such things could exist, and, if they did, how they would interact with the rest of the universe or with each other. They also study even more hypothetical phenomena, such as time travel, hyperspace jumps, and alternate universes.

With the computer, they compare theory with observation, develop theories of their own, and work out advanced calculations that involve light, gravity, time, and the shape of space itself. They also create mathematical computer models.

Some of the phenomena and objects the students study are only the products of wishful thinking. Others are believed to be very real. And others are halfway in between—that is, theory says they should exist, but our descriptions of them, both physical and mathematical, are incomplete or partially in error.

LEARNING ABOUT GALAXIES

Astronomers observe galaxies to try to learn more about how they evolve. To help with this, astronomers and computer scientists have together developed a new system at the Kitt Peak National Observatory called the Interactive Picture-Processing System, IPPS for short. This system uses both traditional photography and special computerized image-enhancement techniques.

Photographs of galaxies are taken and are then digitized and displayed on a TV screen. The IPPS program

The Interactive Picture-Processing System (IPPS) developed at Kitt Peak National Observatory enables astronomers to use and adjust computer enhancement techniques as they observe galaxies, in order to emphasize specific features.

allows the astronomer to type in simple instructions to make the computer change this display to emphasize the features he or she is most interested in. The computer does this by altering the brightness levels of certain portions of the image or by choosing various color-coding schemes—false colors to distinguish between nearly identical areas of brightness or wavelength. In this way, for example, blue stars can be made to stand out from the rest of the galaxy, or those stars that are brighter than a certain value can be specially colored and thus made distinct. Pictures can be produced that are a combination of two or more elements of interest.

The source photographs can be either direct observations already digitized or a pair or set of photographs taken at different wavelengths. The system can also handle images produced by other than optical photography—those created by nonphotographic means.

The computer enables the astronomer to establish the color of the galaxy and the distribution of surface brightness. It can distinguish between red light due to recession and that due to red-giant stars, given the proper instructions.

IPPS is rapid, real-time, and interactive. If one form of enhancement produces no results, another can be tried at once. This increases the efficiency of data analysis, making it possible to study hundreds of galaxies in a variety of ways.

Astronomers generally classify galaxies into types, such as spiral, barred spiral, elliptical, or irregular. Richard B. Larson of Yale University uses a different system of classification—spherical, elliptical, and disk. He has developed complex computer models of these three types. His goal is to discover what galaxies looked like before they assumed their present form. The farther away a galaxy is, the older its light reaching us. Also, the brighter the galaxy was when it was young, the more likely it is that we would be able to see it now. Since spherical galaxies are the brightest, they are the best type to examine to see if they are indeed primeval galaxies. One of the questions Larson hopes to address is whether or not quasars are primeval galaxies.

Peter Quinn of Australian National University used computerized models of spiral galaxies in collision to show that these events can create elliptical galaxies as a consequence. Quinn is also researching, by means of computer models, the observed faint "shell" of stars surrounding some galaxies and has discovered that these shells are present only in elliptical galaxies, not in flat spiral galaxies like the Milky Way. He has also discovered, through his models, that when spiral galaxies collide to form elliptical galaxies, they form shells. Although his conclusions are unproved, of course, they are strong evidence that galaxies often evolve through interactions with each other, and because of the large number of elliptical galaxies that exist today, it can be assumed that such collisions have occurred rather frequently.

CLUSTERS OF GALAXIES

Our galaxy is just one of a group of galaxies, which includes the Great Galaxy in Andromeda. This group is known as the Local Group, and all the galaxies in it, large and small, probably evolved from the same protogalactic mass. We can use models to study how this mass—and we along with it—evolved.

The formation of other clusters of galaxies, some quite far away, can also be studied, frequently using statistical methods. First astronomers have to make an actual count of the number of galaxies that appear on telescopic plates. Then they apply statistical methods to help construct a mathematical model. After that, predictions produced by the model are compared with further observations. When prediction and observation match, we know that our model represents reality—more or less.

X rays are especially useful for examining distant galaxies and clusters of galaxies. One particular distant galaxy, the main member of a group, is M87, which has an X-ray halo surrounding it. A computer representation, in false colors, indicates that M87 has greater mass than we can detect optically. This extra mass helps hold the related cluster of galaxies together. Discovering this unseen mass has helped us begin to answer questions about the evolution of galaxies and the total mass of the universe.

From there, we can study and speculate on how the universe itself evolved. Knowing what we do (and what we don't) about the Big Bang, we can begin to formulate questions about whether the universe is open or closed and whether it will die when all the energy in it is evenly distributed or gravity will be strong enough to pull everything back together to start things over again.

THE SHAPE AND FUTURE OF THE UNIVERSE

Models of the structure of the universe let us look back in time to the Big Bang, to test hypotheses of how the universe evolved from the beginning, what conditions were like then, and how they became what they are now. Even in such a theoretical study as this there are possible practical consequences, as we discover the true nature of matter and energy. Mathematical theories concerning the nature of the universe, drawing on relativity and other branches of physics, give us some idea of how the universe began.

A computer model created by Edward J. Groth, P. James E. Peebles, Michael Seldner, and Raymond M. Soneira, cosmologists at Princeton University, studies the distribution of galaxies in the universe. The model was made after extensive statistical analyses were done on galaxies appearing on telescopic plates. When the distribution of the model matches the distribution observed, then the mathematics of the model will be confirmed. This model can then be used as the basis for a model of the evolution of clusters of galaxies and of the universe itself.

Astronomers want to learn not only how galaxies formed but also how they will evolve in the future. Once they have a good model of the history of the universe, or even of parts of it, they can run it forward, past the present time, to extrapolate what the cosmos will be like millions or billions of years from now. This is to help them determine how long the universe will last, and what changes it will go through in time.

One of the as yet unanswered questions is whether the universe is open or closed. That is, will it expand forever or eventually recondense perhaps to form another Big Bang? In order to choose between an open or closed possibility, astronomers compare their models with observable features of the real universe of today. They hope to find evidence that will confirm one of the possible futures of the universe and exclude all others.

Our knowledge is still pretty limited in this area. So far there has been no single observation or test or measurement precise enough to settle the question. Some evidence leads us to assume that the universe is open, but new discoveries hint that this might not be true. The models the astronomers use suggest further tests, different observations, and new areas of exploration. Someday, we may know for sure.

INDEX